Aims and Scope

The series "Springer Theses" brings together a selection of the very best Ph.D. theses from around the world and across the physical sciences. Nominated and endorsed by two recognized specialists, each published volume has been selected for its scientific excellence and the high impact of its contents for the pertinent field of research. For greater accessibility to non-specialists, the published versions include an extended introduction, as well as a foreword by the student's supervisor explaining the special relevance of the work for the field. As a whole, the series will provide a valuable resource both for newcomers to the research fields described, and for other scientists seeking detailed background information on special questions. Finally, it provides an accredited documentation of the valuable contributions made by today's younger generation of scientists.

Theses are accepted into the series by invited nomination only and must fulfill all of the following criteria

- They must be written in good English.
- The topic should fall within the confines of Chemistry, Physics, Earth Sciences, Engineering and related interdisciplinary fields such as Materials, Nanoscience, Chemical Engineering, Complex Systems and Biophysics.
- The work reported in the thesis must represent a significant scientific advance.
- If the thesis includes previously published material, permission to reproduce this must be gained from the respective copyright holder.
- They must have been examined and passed during the 12 months prior to nomination.
- Each thesis should include a foreword by the supervisor outlining the significance of its content.
- The theses should have a clearly defined structure including an introduction accessible to scientists not expert in that particular field.

More information about this series at http://www.springer.com/series/8790

Diane Ansermet

Emergent Superconductivity in Low Dimensions

Doctoral Thesis accepted by
the Nanyang Technological University,
Singapore, Singapore

Springer

Author
Dr. Diane Ansermet
Division of Physics and Applied Physics, School of Physical and Mathematical Sciences
Nanyang Technological University
Singapore, Singapore

Supervisor
Prof. Christos Panagopoulos
Division of Physics and Applied Physics, School of Physical and Mathematical Sciences
Nanyang Technological University
Singapore, Singapore

ISSN 2190-5053 ISSN 2190-5061 (electronic)
Springer Theses
ISBN 978-981-13-2940-1 ISBN 978-981-13-2941-8 (eBook)
https://doi.org/10.1007/978-981-13-2941-8

Library of Congress Control Number: 2018958344

This Springer imprint is published by the registered company Springer Nature Singapore Pte Ltd.
The registered company address is: 152 Beach Road, #21-01/04 Gateway East, Singapore 189721, Singapore

Pour toutes les adolescentes qui un jour
ont cru detester la physique

Supervisor's Foreword

Diane's thesis unites a variety of very delicate experimental techniques with supporting calculations to address collectively a wide range of interrelated themes in condensed matter physics: strong electronic correlations, disorder and inhomogeneity, low dimensionality, localization, and superconductivity. Her work is inspirational for materials engineers as well as fundamental scientists.

Her discovery that superconducting phase coherence can be recovered ("re-entrant") in low-dimensional materials subjected to high temperatures, magnetic fields, and electric currents suggests that artificial superconducting nano-composites are realistically promising for technological applications.

Furthermore, her study of 1D–3D crossover in the electronic properties of both the superconducting and metallic states is already inviting deeper theoretical attention from the academic community.

Perhaps, the most profound contribution from Diane's research is the observation of disorder-enhanced superconductivity due to intrinsically screened Coulomb interaction in disordered quasi-1D systems. This discovery suggests that other inhomogeneous quasi-1D metals may also display anomalously correlated superconducting and/or magnetic ground states, thus opening an entirely new avenue of research in low-dimensional electronic materials.

Diane's thesis provides an outstanding overview of the "state of the art" in low-dimensional superconductivity and a promising vision of the future.

Singapore
July 2018

Prof. Christos Panagopoulos

Abstract

Low-dimensional systems provide the opportunity to explore the relationship between electronic correlations, dimensionality, inhomogeneities, and superconductivity. In this work, single crystals of the quasi-one-dimensional $Na_{2-\delta}Mo_6Se_6$ are studied, composed of MoSe filaments weakly coupled by Na atoms and subject to intrinsic disorder ($\delta > 0$). $Na_{2-\delta}Mo_6Se_6$ is demonstrated to display strong electronic correlations in its normal state, whereas a superconducting ground state emerges from Anderson localized electrons. Two novel behaviors of the superconducting state are observed: first, a disorder-induced enhancement of the superconducting transition temperature; second, a re-entrant phase coherence with increasing temperature, magnetic field, and current. The intrinsic properties of $Na_{2-\delta}Mo_6Se_6$ are analyzed to offer a thorough understanding of these phenomena. The emergence of superconductivity in such low-dimensional systems provides a fruitful playground to explore electronic order and correlations.

Parts of this thesis have been published in the following journal articles:

[1] D. Ansermet ★, A. P. Petrović ★, S. He, D. Chernyshov, M. Hoesch, D. Salloum, P. Gougeon, M. Potel, L. Boeri, O. K. Andersen, C. Panagopoulos. Reentrant phase coherence in superconducting nanowire composites. *ACS Nano* **2016**, *10*, 515–525.

[2] A. P. Petrović ★, D. Ansermet ★, D. Chernyshov, M. Hoesch, D. Salloum, P. Gougeon, M. Potel, L. Boeri, C. Panagopoulos. A disorder-enhanced quasi-one-dimensional superconductor. *Nature Communications* **2016**, *7*, 12262.

Declaration of Authorship

I, Diane Ansermet, declare that this thesis titled, 'Emergent Superconductivity in Low Dimensions' and the work presented in it are my own. I confirm that:

- This work was done wholly or mainly while in candidature for a research degree at this University.
- Where any part of this thesis has previously been submitted for a degree or any other qualification at this University or any other institution, this has been clearly stated.
- Where I have consulted the published work of others, this is always clearly attributed.
- Where I have quoted from the work of others, the source is always given. With the exception of such quotations, this thesis is entirely my own work.
- I have acknowledged all main sources of help.
- Where the thesis is based on work done by myself jointly with others, I have made clear exactly what was done by others and what I have contributed myself.

February 2017

Diane Ansermet

Acknowledgements

The work presented in this thesis was performed in the laboratory of Prof. C. Panagopoulos, Physics of Novel Electronics (PHYNE), in NTU. All experiments were supported by the National Research Foundation, Singapore, through Grant NRF-CRP4-2008-04. My scholarship was funded partly by a Tier 2 grant and partly by a NTU Research Scholarship. I am very grateful for these financial supports.

I would first like to warmly thank my supervisor, Prof. Christos Panagopoulos. Following my bachelor's and master's degrees in Material Sciences in EPFL, Switzerland, he gave me this chance to do a Ph.D. in Applied Physics. This allowed me to learn a set of highly skilled measurement techniques and study a field completely new to me. He taught me to surpass my own expectations and always seek greater achievements.

I could not have succeeded without the continuous support and teaching of Alexander P. Petrović. He taught me (almost!) everything I know about superconductivity. We went through this fantastic journey of discovering $Na_{2-\delta}Mo_6Se_6$ together, and his enthusiasm kept me going. He showed me how experimental physics can be a fun game of hide and seek with data analysis. Alex: it was fun, indeed!

My laboratory mates Sai Swaroop Sunku, Xian Yang Tee, Shikun He, and Anjan Soumyanarayanan made the experimental learning experience highly memorable, and I will cherish the good (and less good) times we shared installing, fixing, and operating the different systems. During the long measurements periods, the laboratory felt like a little family, and I thank them all for that. I would also like to thank my other colleagues, Antoine Paré, Irene Neo Ai Ling, M. Raju, Bhartendu Satywali, Sreemanta Mitra, and Chern Hui Lee. All of them taught me valuable lessons of their own, and I grew thanks to them.

Our collaborators cited along this thesis provided valuable data and calculations to complete this study, and I am extremely thankful for their quality work.

On a more personal level, Hunter: I love you!!! Our joined trip into this Ph.D. journey was one of a kind. You showed me unconditional love and support through the years and reminded me to keep believing when I was loosing myself. Without you, none of this would have been possible.

To all my Singapore friends and the ones back at home who reminded me to live a little, thank you! A special note to Chris: Our regular coffee breaks were a little ray of sunshine keeping me alive in the laboratory. And of course, to my family: You believed in me, supported my choices, came to visit and discover this new life. All these have been priceless.

Contents

Abbreviations

1D	One-dimensional
2D	Two-dimensional
3D	Three-dimensional
AC	Alternative current
ACT	AC transport
BCS	Bardeen–Cooper–Schrieffer
BKT	Berezinski–Kosterlitz–Thouless
BZ	Brillouin zone
CDW/SDW	Charge/spin density wave
CHE	Continuous heat exchanger
DC	Direct current
DFT	Density functional theory
DoS	Density of states
e-e	Electron-electron
EM	Electromagnetic
e-ph	Electron-phonon
FC (ZFC)	Field cooled (zero)
FFLO	Fulde–Ferrell–Larkin–Ovchinnikov
GL	Ginzburg–Landau
ICP	Intermediate cold plate
IR	Infrared
LHe	Liquid helium
LL	Luttinger liquid
LN_2	Liquid nitrogen
MC	Mixing chamber
MCU	Mixture condensing unit
MI	Metal-insulator
MPMS (PPMS)	Magnetic (physical) property measurement system
OFHC	Oxygen-free high thermal conductivity
OVC	Outer vacuum chamber

psd	Phase-sensitive detection
q1D	Quasi-one-dimensional
QPS	Quantum phase slips
RF	Radio frequency
RTD	Resistance temperature detector
SEM	Scanning electron microscope
SQUID	Superconducting quantum interference device
TAPS	Thermally activated phase slips
TDS	Thermal diffuse scattering
VRH	Variable range hopping
VTI	Variable temperature insert
WHH	Werthamer–Helfand–Hohenberg

Symbols

E_a	Activation energy for Josephson tunneling (eV)
E_c	Localization critical energy (=mobility edge) (eV)
E_g	Energy gap (eV)
E_F	Fermi energy (eV)
E_J	Josephson energy (eV)
H_c (H_{c2})	Critical field (in type-II superconductor) (T (= 10^4 Oe))
H_P	Pauli limit (T (= 10^4 Oe))
I_c	Critical current (A)
l	Electron mean free path (m)
j_s	Superconducting current density (A/m^2)
m	Electron pair mass (= $2m_e$) (kg)
$\mathbf{k}$ ($\mathbf{k}_F$)	Wave vector (at the Fermi energy) (1/m)
$N(E)$	Electron density of states (= DoS) (1/eV)
n_s	Local density of superconducting electrons (−)
T_{BKT}	BKT transition temperature (K)
T_J	Josephson temperature (K)
T_{MI}	Metal-insulator transition temperature (K)
T_c	Critical temperature (K)
T_{ons}	Onset temperature of superconducting fluctuations (K)
T_{pin}	Charge pinning temperature (K)
T_R	Re-entrance threshold temperature (K)
T_x	Single-particle dimensional crossover temperature (K)
$t_{\parallel}$	Longitudinal hopping energy (eV)
$t_{\perp}$	Transverse hopping energy (eV)
R_N	Normal state resistance at T_{ons} (Ω)
R_n	Normal state resistance of a 1D wire (Ω)
R_t	Tunneling resistance (Ω)
U	Coulomb repulsion energy (eV)
Δ	Superconducting energy gap (eV)
ϵ_k	Fermi level (eV)

K_ρ	Luttinger charge parameter (−)
λ_L	London penetration depth (m)
ν_F	Fermi velocity (m/s)
ξ_L	Localization length (m)
$\xi(T)$	GL coherence length (m)
ξ_0	BCS coherence length (m)
ρ	Electrical resistivity (Ωm)
σ	Electrical conductivity (S/m)
$\Psi(r)$	Superconducting complex order parameter (−)
ω_D	Debye frequency (Hz)

List of Figures

List of Tables

Physical Constants

Bohr magneton	$\mu_B = e\hbar/2m_e c \approx 9.274 \times 10^{-24}$ J/T
Boltzmann constant	$k_B \approx 1.381 \times 10^{-23}$ J/K
Electron charge	$e \approx 1.602 \times 10^{-19}$ C
Electron mass	$m_e \approx 9.109 \times 10^{-31}$ kg
Magnetic flux quantum	$\Phi_0 \equiv \frac{h}{2e} \approx 2.067 \times 10^{-15}$ Wb
Planck's constant	$h \approx 6.626 \times 10^{-34}$ Js
	$\hbar = h/2\pi \approx 1.054 \times 10^{-34}$ Js
Speed of light	$c \approx 2.997 \times 10^8$ m/s
Vacuum permeability	$\mu_0 = 4\pi \times 10^{-7}$ N/A^2

Chapter 1
Superconductivity: History and Motivations

1.1 History

In 1911 in Leiden, H. K. Onnes studied the transport properties of mercury using liquid helium, and saw its electrical resistivity suddenly disappear [1]. He observed this phenomenon in several other elements as well and gave it the name "superconductivity". Approximately 20 years later, in 1933, W. Meissner and R. Ochsenfeld [2] showed that a material in the superconducting state expels the applied external magnetic field, becoming a perfect diamagnet. This behavior is now commonly known as the Meissner effect. These discoveries settled the two distinctive characteristics of a superconductor: below a critical temperature, it shows perfect electrical conductivity and diamagnetism.

It was first the brothers F. and H. London in 1935 who provided a theoretical understanding of the superconducting state [3]. They explained how the minimization of the free energy carried by the supercurrent was responsible for the Meissner effect. They developed phenomenological models which were formulated along the fundamental observations but did not explain how they were connected to each other.

Then in 1950, V. L. Ginzburg and L. Landau (GL) proposed a phenomenological theory [4] which successfully explained the macroscopic properties of superconductors. They introduced the superconducting order parameter which describes the superconducting electrons. The same year, the dependence of the superconducting critical temperature to the isotopic mass of the material was observed [5], which suggested that superconductivity might be taking place through electron-phonon interactions. Although a microscopic understanding was still missing, GL theory allowed A. A. Abrikosov to predict two categories of superconductors [6], known today as type-I and type-II.

A complete understanding of the microscopic mechanism was finally proposed by J. Bardeen, L.N. Cooper and J.R. Schrieffer (BCS) in 1957 [7]. The BCS theory explained how electrons can reduce their energy by pairing mediated by phonons (Cooper pairs), resulting in a gap in the density of states at the Fermi energy. This theory revolutionized the field of superconductivity and is still nowadays the most

D. Ansermet, *Emergent Superconductivity in Low Dimensions*, Springer Theses,
https://doi.org/10.1007/978-981-13-2941-8_1

advanced understanding we have of the superconducting state. A few years later, L. Gor'kov showed how GL theory was a limited form of the BCS theory [8] and how the superconducting order parameter was proportional to the energy of the electron pairs.

The outcome of the GL and BCS theories allowed the field of superconductivity to advance at a great pace: alloyed materials were developed into commercial superconducting wires in 1962. In the same year, Josephson discovered that supercurrents could tunnel through a thin insulator [9] which led to further commercial applications such as superconducting quantum interference devices (SQUIDs). However, the critical temperature of all known superconductors still remained incredibly low ($<< 77$ K, liquid nitrogen temperature).

In 1986, the discovery of a critical temperature at 35 K in the perovskite lanthanum barium cooper oxide (LBCO) was made by G. Bednorz and K.A. Müller in the IBM laboratory in Zürich [10]. Shortly after, higher critical temperatures were discovered in many similar materials [11], and reached temperatures above the boiling point of nitrogen. The microscopic mechanism responsible for their superconducting state can not be fully explained by the BCS theory and these materials provided new areas and challenges to explore high temperature superconductivity.

Today, we still stand far from the holy grail of room temperature superconductivity [12, 13]. The high-temperature superconductors still lack complete theoretical understanding and face several experimental barriers. The field of superconductivity seems to need a fresh, new approach to looking for functional superconductors, within maybe some forgotten, but fruitful materials.

1.2 Motivations: A Fresh Look Towards Functional Superconductors

The motivations for this work rose from, first, the desire to understand better the establishment of superconductivity in low-dimensional materials. The aim was to comprehend the limits of superconducting fluctuations as they are detrimental to the stability of the superconducting state. As fluctuations are inherently present in one-dimensional (1D) superconductors, low-dimensional materials are ideal candidates for their study.

Secondly, low dimensions provide the opportunity to explore the relationship between superconductivity, electronic correlations, dimensionality, and disorder. In order to develop new functional superconductors, it is crucial to include aspects of real materials (e.g. disorder, dimensionality) in the investigation. The objectives were to explore the influence of the normal state on the emergence and stability of a superconducting state.

It is generally acknowledged that disorder impacts strongly 1D materials compared to 3D systems. However, the effect of disorder on 3D systems subject to strong electronic correlations is still poorly understood: usually, electronic properties in such systems are overly complicated to simulate and experimental data are difficult to interpret.

The study of quasi-1D (q1D) systems can provide insightful information to the field of strongly correlated materials. A q1D material consists of 1D nanofilaments weakly coupled perpendicular to their main axis. The coupling of a multitude of filaments forms a 3D matrix whose electronic properties are still highly influenced by the system's anisotropy. As 1D physics is relatively well understood, one would "simply" need to add weak coupling between the filaments to simulate the bulk properties. In addition, if disorder is introduced into the matrix, all the ingredients are in hand: strong electronic correlations, disorder and, with the right choice of material, superconductivity as well.

The key aims driving applied superconductivity are as follows: first, developing functional superconductors that can be used at higher temperature, magnetic field or current. Secondly, modeling new materials resilient to inhomogeneity (disorder) and superconducting fluctuations (destruction of coherent superconductivity). However, both aims seem to be incompatible since increasing temperature, magnetic field, current, or disorder are inherently causing fluctuations and destruction of the superconducting state.

The stability of the superconducting state with dimensionality and disorder can be explored using a q1D material: theories developed for 1D systems can easily be adapted to the q1D morphology, and strong electronic correlations and disorder present in the normal state can provide information about the emergence of the superconducting state. Unfortunately, fabrication techniques for nanofilamentary composites are still in their infancy, limiting the choice of materials for such study.

In consequence, the present work focuses on q1D bulk single crystals exhibiting intrinsic disorder: $Na_{2-\delta}Mo_6Se_6$ where $\delta > 0$ represents Na vacancies. This investigation provides the very first experimental analysis of their electronic ground state and its interplay with disorder and dimensionality.

The scope of the results presented in this thesis is beyond the specific material of this study: it highlights universal behaviors and provides information on disordered q1D systems. The interdependence between the nature of the normal and the superconducting states show promises for future applications. This work is anticipated to instigate the development of new engineered nanofilamentary materials exhibiting similar or enhanced superconducting properties.

References

1. H.K. Onnes. Further experiments with liquid helium. In Kostas Gavroglu and Yorgos Goudaroulis (eds.), *Through Measurement to Knowledge: The Selected Papers of Heike Kamerlingh Onnes 1853–1926*, (Springer Netherlands, Dordrecht, 1991) pp. 201–220
2. W. Meissner, R. Ochsenfeld, Ein neuer Effekt bei Eintritt der Supraleitfähigkeit. Naturwissenschaften **21**(44), 787–788 (1933)

3. F. London, H. London, The electromagnetic equations of the supraconductor. Proc. R. Soc. A: Math. Phys. Eng. Sci. **149**(866), 71–88 (1935)
4. L.D. Landau, D. Ter-Haar, *Collected papers of L.D. Landau* (Pergamon Press, Oxford, 1965)
5. E. Maxwell, Isotope effect in the superconductivity. Phys. Rev. **78**, 477 (1950)
6. A.A. Abrikosov, Magnetic properties of superconductors of the second group. Sov. Phys. JETP **5**(6) (1957)
7. J. Bardeen, L.N. Cooper, J.R. Schrieffer, Theory of superconductivity. Phys. Rev. **108**(5), 1175–1204 (1957)
8. L. Gor'kov, Microscopic derivation of the Ginzburg-Landau equations in the theory of superconductivity. Sov. Phys. JETP **36**(9), 1364–1367 (1959)
9. B.D. Josephson, Possible new effects in superconductive tunnelling. Phys. Lett. **1**(July), 7–9 (1962)
10. J.G. Bednorz, K.A. Müller, Possible high Tc superconductivity in the Ba-La-Cu-O system. Z. für Phys. B Condens. Matter **193**, 189–193 (1986)
11. A.S. Davydov, Theoretical investigation of high-temperature superconductivity. Phys. Rep. **4 and 5**(190), 191–306 (1990)
12. A. Mourachkine, *Room-Temperature Superconductivity* (Cambridge Int Science Publishing, 2006)
13. Alex Gurevich, To use or not to use cool superconductors? Nat. Mater. **10**(4), 255–259 (2011)

Chapter 2
Introduction and Theory

2.1 Superconductivity

2.1.1 *Experimental Signatures of Superconductivity*

When a superconducting material is cooled below its critical temperature T_c, it undergoes an electronic phase transition wherein the electrons pair to form a boson with integer spin. This new particle, called Cooper pair, can move in the crystal without dissipation. This leads to the first signature of superconductivity: below T_c, the electrical resistance drops to zero (Fig. 2.1a). Additionally, in the presence of an applied magnetic field, the superconducting material establishes persistent electric currents near its surface, which induce an expulsion of all magnetic flux (perfect diamagnetism). This leads to the second signature of superconductivity: the Meissner effect (Fig. 2.1b).

2.1.2 *A Macroscopic Quantum Phenomenon*

Although every atom in a macroscopic object obeys the law of quantum mechanics, it is generally not possible to observe quantum effects on the macroscopic scale: thermal motion masks quantum irregularities. However, in the case of superconductivity, it is possible to witness macroscopic quantum phenomena: the parameters that characterize macroscopic objects (e.g. the flux in a superconducting ring) are quantized. In a homogeneous superconductor, the electron system is highly coherent: the superconducting electrons have to be considered as a single quantum mechanical entity. This was first observed and developed in the phenomenological theories of the London brothers [2].

D. Ansermet, *Emergent Superconductivity in Low Dimensions*, Springer Theses,
https://doi.org/10.1007/978-981-13-2941-8_2

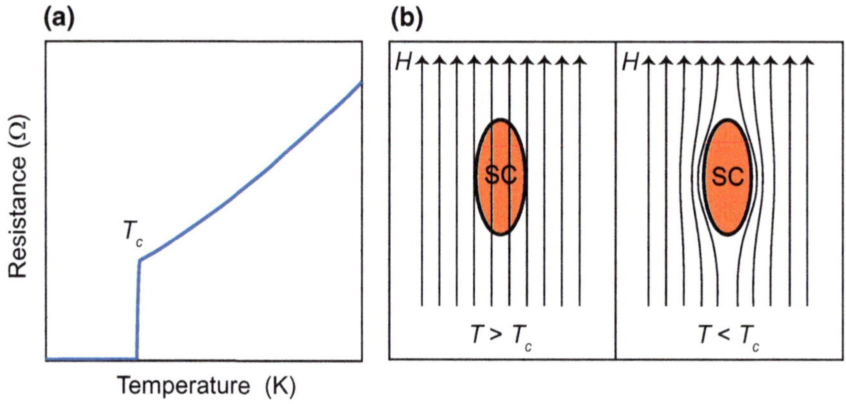

Fig. 2.1 Schematic of the two signatures of superconductivity: **a** The resistance drops to zero below the critical temperature T_C. **b** The Meissner effect, where the superconductor (SC) exhibits perfect diamagnetism at $T < T_C$. Inspired from Tinkham [1]

The London Equations

The first London equation describes the perfect conductivity of the superconducting state (Fig. 2.1a). A superconductor preserves the original current distribution within the bulk limit:

$$\frac{\delta j_s}{\delta t} = \frac{n_s e^2}{m}\mathbf{E} \tag{2.1}$$

where j_s is the superconducting current density, n_s the local density of superconducting electrons and **E** the electric field.

The second London equation describes the Meissner effect and illustrates the expulsion of the magnetic field **B** within a superconductor:

$$\nabla \times j_s = \frac{n_s e^2}{m}\mathbf{B} \tag{2.2}$$

These equations were developed to explain the experimental observations of the superconducting transition. However, they do not explain how these phenomena are linked to each other. While providing a first glimpse of the macroscopic scale of the superconducting state, they gave key concepts for the complete understanding of superconductivity developed later.

Flux Quantization

A superconducting ring has a stable supercurrent due to the zero resistance state and is therefore in a stationary state. As superconductivity is a macroscopic quantum phenomenon, the stationary state is determined following quantum conditions: the electron states are controlled by the requirement that the wavefunction around the loop be single valued, so the phase has to wind by 2π. This imposes a quantization condition on the superconducting state: the magnetic flux inside a superconducting ring can only have discrete values of the magnetic flux quantum Φ_0:

$$\Phi = n\Phi_0 \tag{2.3}$$

where n is an integer and $\Phi_0 \equiv \frac{h}{2e} \approx 2.07 \times 10^{-15}$ Wb.

This was first observed by Fritz London [3] who wrongly calculated the value of the flux quantum considering that single electrons were carrying the supercurrent. This will be corrected only with the BCS theory, describing the superconducting carrier as a pair of electrons.

2.1.3 Bardeen–Cooper–Schrieffer Theory

The experimental observation of a dependence of the value of T_c with the mass of the material's isotopes (isotope effect) suggested that phonons were involved in the electron pairing mechanism. This was considered by J. Bardeen et al. [4] who formulated the BCS theory, transforming our understanding of superconductivity.

In a superconductor at $T = 0$, electrons sufficiently near the Fermi level may couple, forming Cooper pairs. To understand the formation mechanism, the motion of the ions in a material needs to be considered. Electrons move with Fermi velocity $\nu_F = \hbar k_F/m_e$, while ions move with a much slower velocity $\nu_i = \nu_F m_e/m_i$ where m_e and m_i are the electron and ion mass, respectively. A flowing electron will attract positive ions and therefore polarize their environment. As $\nu_i << \nu_F$, this polarization takes a long time ($\tau \sim 2\pi/\omega_D \sim 10^{-13}$ s, ω_D being the Debye frequency) which will provoke an attraction on a second flowing electron. This will lead to an effective positive interaction between the two electrons (Fig. 2.2), with a distance $\nu_F \cdot \tau \sim$ 100 nm. If this effect is larger than any present repulsive Coulomb interactions, this will result in a net attraction between the two electrons, and the Cooper pair will be the most energetically favorable state.

The polarization induces a lattice deformation, which is characterized by the vibrational (i.e. phonon) frequencies. To conserve the momentum during the electron-electron (e-e) interaction, the resulting phonon must have a frequency lower than a critical value ω_D. This defines the cut-off energy $\hbar\omega_D$ of the pair's attractive potential. If the electrons energy differences are higher than $\hbar\omega_D$, the interaction becomes repulsive. The pairing energy is therefore directly dependent on the mass M of the lattice ions, as $\omega_D \propto M^{-1/2}$, thus explaining the isotope effect.

The critical temperature T_c at which the superconducting state is unstable is:

$$T_c = 1.13\frac{\hbar\omega_D}{k_B}\exp\left(-\frac{1}{N(0)V}\right) \tag{2.4}$$

Which can be approximated as:

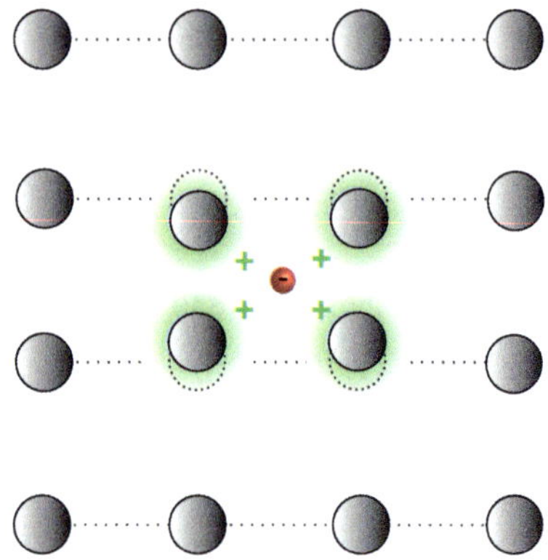

Fig. 2.2 An electron creates a polarization of the lattice structure due to the positive charge of the ions. This will in turn attract another electron, creating a net attraction between the two electrons

$$T_c \propto \omega_D \exp\left(-\frac{1}{\lambda_{e-ph}}\right) \tag{2.5}$$

where $\lambda_{e-ph} = N(0)V$ is the electron-phonon interaction with $N(0)$ the density of states (DoS) at the Fermi level for electrons of one spin orientation and V the interaction potential.

Cooper pairs are usually formed by two electrons with opposite spin and momentum.[1] They behave as single particles (bosons) and are characterized by a wavefunction obtained from the two particle wavefunction with the lowest-energy state with zero total momentum:

$$\psi(\mathbf{r}_1 - \mathbf{r}_2) = \sum_{\mathbf{k}>\mathbf{k_F}} g_{\mathbf{k}} e^{i\mathbf{k}(\mathbf{r}_1-\mathbf{r}_2)} \tag{2.6}$$

where $\mathbf{r}_1$ and $\mathbf{r}_2$ are the positions of the two electrons in real space and $g_{\mathbf{k}}$ corresponds to the probability to find the Cooper pair in real space. ψ is complex with an amplitude and a phase.

The energy separating the ground state from the excited states is known as the pairing energy Δ, i.e. the superconducting energy gap. The electron energy relative to the Fermi level $\epsilon_{\mathbf{k}}$ is dependent on the electron momentum $\mathbf{k}$. These two terms describe the elementary excitation energies of a superconducting system as follows:

$$E_{\mathbf{k}} = \sqrt{\epsilon_{\mathbf{k}}^2 + |\Delta^2|} \tag{2.7}$$

where $E_{\mathbf{k}}$ is the excitation energy of a quasiparticle of momentum $\hbar\mathbf{k}$, known as the Bogoliubov quasiparticle [1].

[1] Some exceptions occur: for example Fulde–Ferrell–Larkin–Ovchinnikov (FFLO) phases emerge from Cooper pairs with non-zero net momentum.

At the Fermi surface where $\epsilon_{\mathbf{k}} = 0$, $E_{\mathbf{k}} = \sqrt{|\Delta^2|} > 0$ and provides an energy scale for the stability of the superconducting state. Therefore, 2Δ can be understood as the binding energy of a Cooper pair.

The energy gap is expected to reach zero at T_c (as the superconducting state is destroyed above T_c), leading to $E_{\mathbf{k}} \rightarrow |\epsilon_{\mathbf{k}}|$. Hence, at $T > T_c$, the excitation energy is equal to that of an electron in the normal state.

Using this definition together with the symmetry of $|\epsilon_{\mathbf{k}}|$ and the derivations of the matrix elements of the interaction potential $V_{\mathbf{kk}'}$ (representing the scattering of the Cooper pair with momentum $(\mathbf{k}, -\mathbf{k})$ to $(\mathbf{k}', -\mathbf{k}')$), a useful estimate for the energy gap at $T = 0$ is obtained [1]:

$$2\Delta_0 \approx 3.5kT_c \tag{2.8}$$

For $0 < T < T_c$, the energy gap is expected to decrease with increasing temperature (Fig. 2.3).

The gap is weakly affected by the increase of T for $T << T_c$, until a significant number of particles become thermally excited: close to T_c, the gap magnitude is approximated as follows:

$$\frac{\Delta(T)}{\Delta_0} \approx 1.76\left(1 - \frac{T}{T_c}\right)^{1/2} \tag{2.9}$$

The characteristic size of a Cooper pair is described as the BCS coherence length ξ_0:

$$\xi_0 = \frac{\hbar v_F}{\pi \Delta_0} \tag{2.10}$$

The BCS approximations remain valid in the weak-coupling limit $\lambda_{e-ph} = N(0)V \ll 1$. Except when mentioned, the experimental results and analysis presented in this work were considered in the weak-coupling limit.

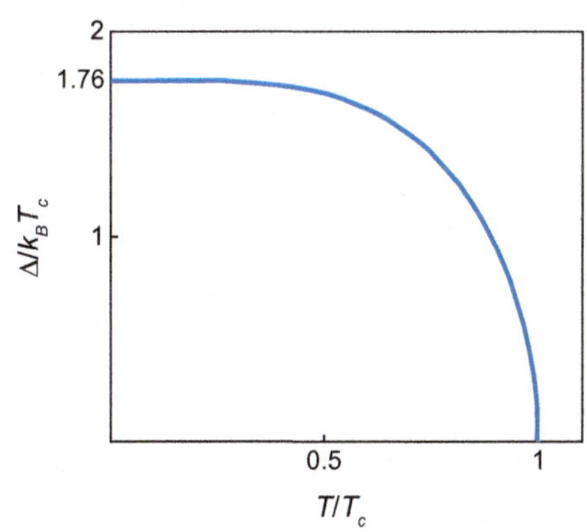

Fig. 2.3 Evolution of the BCS energy gap Δ as a function of temperature T close to T_c, following Eq. 2.9

2.1.4 Ginzburg–Landau Theory: the Complex Order Parameter

While BCS theory provides a detailed microscopic understanding of the superconducting state, it is also necessary to consider macroscopic variations such as spatial inhomogeneities. Ginzburg–Landau (GL) theory [5] provides a phenomenological framework to consider the spatial evolution of superconductivity.

GL theory introduces the superconducting condensate wavefunction in the form of a complex order parameter $\Psi(\mathbf{r})$:

$$\Psi(\mathbf{r}) = |\Psi(\mathbf{r})|\, e^{i\varphi(\mathbf{r})} \tag{2.11}$$

where the amplitude $|\Psi(\mathbf{r})|^2$ represents the local density of superconducting electrons $n_s(\mathbf{r})$ and the phase φ gradient determines the supercurrent flow. Any variations of $n_s(\mathbf{r})$ (and therefore $|\Psi(\mathbf{r})|$) are proportional to the variation in the BCS energy gap Δ, i.e. the binding energy of the electron pairs.

In cases where $n_s(\mathbf{r})$ varies spatially, the theory considers an expansion of the free-energy density f in powers of $|\Psi|^2$ [1]. This expansion assumes that if $\Psi = 0$ the free energy density is that of the normal state f_{ns}:

$$f(|\Psi|^2, T) = f_{ns} + \alpha\,|\Psi|^2 + \frac{1}{2}\beta\,|\Psi|^4 + \cdots \tag{2.12}$$

where α and β are temperature dependent phenomenological parameters of the theory. By definition, the superconducting state is stable when the free energy is minimized, i.e. when $\mathrm{d}f/\mathrm{d}|\Psi|^2 = 0$.

This leads to the following for $T < T_c$:

$$|\Psi|^2 = -\alpha/\beta \tag{2.13}$$

The evolution of the complex order parameter can then be traced by minimizing f when subject to small spatial variations.

The GL theory defines two characteristic lengths in a superconductor:
1. The GL penetration depth λ which corresponds to the depth of the penetration of an external magnetic field inside the superconductor:

$$\lambda = \sqrt{\frac{mc^2\beta}{8\pi e^2\,|\alpha|}} \tag{2.14}$$

where $m = 2m_e$ is the Cooper pair mass.

This was first derived by the London brothers (from Eq. 2.2) using Ampere's law.[2]
2. The GL coherence length ξ which corresponds to the characteristic length over which spatial variations in Ψ take place:

[2] Ampere's law: $\nabla \times \mathbf{B} = \mu_0 \mathbf{J}$ where $\mathbf{J}$ is the current density.

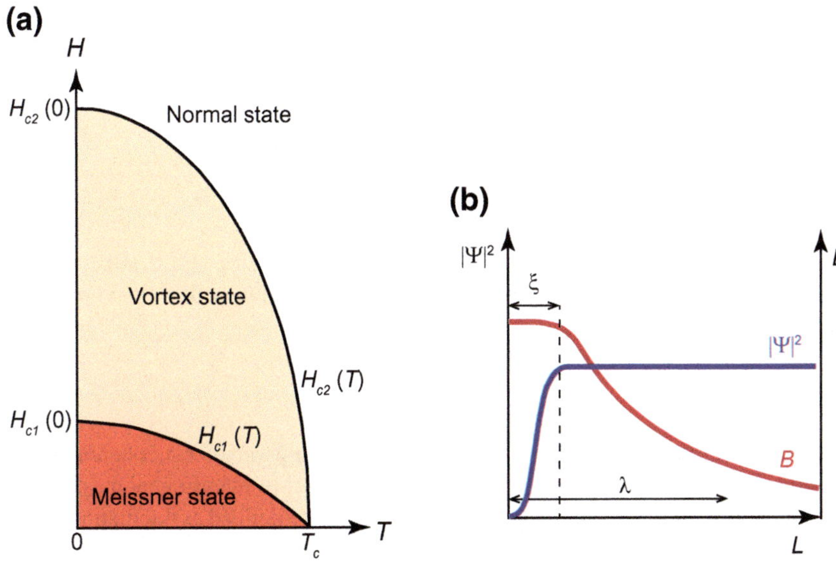

Fig. 2.4 Schematics of **a** the temperature dependence of the critical fields H_{c1} and H_{c2} in a type-II superconductor, and **b** the spatial variation of the amplitude of the order parameter $|\Psi|^2$ and the magnetic field B inside a vortex in a type-II superconductor. L is the distance from the center of the vortex. Adapted from Mourachkine [7]

$$\xi = \frac{\hbar}{\sqrt{2m\,|\alpha|}} \tag{2.15}$$

The GL penetration and coherence lengths show the same dependence to $\alpha \propto (T - T_c)$ near T_c which diverges as $(T_c - T)^{-\frac{1}{2}}$ at $T \sim T_c$. Their ratio $\kappa = \lambda/\xi$ provides a useful tool to describe superconductors: type-I for $\kappa < 1/\sqrt(2)$ and type-II for $\kappa > 1/\sqrt(2)$ have been defined due to their very different behavior in a magnetic field [6].

Type-I superconductors fully expel all external magnetic field at $T < T_c$, leading to a single critical field $H_c(0)$.

In type-II superconductors, the field first partially penetrates the material at $H_{c1}(0)$ in the form of vortices, and then fully destroys superconductivity at $H_{c2}(0)$ (Fig. 2.4a). In the vortex state (between $H_{c1}(0)$ and $H_{c2}(0)$), the magnetic field is large inside the vortices, exposing a normal state, while the superconducting state is maintained everywhere else.

Selecting an isolated vortex, the physical meaning of ξ and λ for the spatial variation of $|\Psi|^2$ and in applied field B is presented in Fig. 2.4b.

Type-I superconductivity is present in most bulk metals such as aluminum, lead or mercury, while type-II superconductors are usually alloys or complex oxide ceramics (cuprates).

The difference between the GL coherence length ξ and the BCS coherence length ξ_0 can be understood as follows: ξ_0 is the Cooper pair size at $T \sim 0$ [1], whereas ξ is temperature dependent and provides a range over which the phase of the order parameter remains stable:

$$\xi(T) = \frac{\xi_0}{\sqrt{1 - T/T_c}} \tag{2.16}$$

Thus, ξ can be calculated directly from the BCS expression (Eq. 2.10).

In the framework of the GL theory, the flux quantization within the vortex core leads to an important relationship between H_{c2} and ξ:

$$\Phi_0 = 2\pi\xi^2 H_{c2} \tag{2.17}$$

In ideal bulk superconductors, the phase of the order parameter is coherent everywhere at $T < T_c$. However, in superconductors subject to inhomogeneities or with reduced dimensions, the phase may be incoherent: this is defined as fluctuating superconductivity.

A true zero resistance will arise only from complete phase coherence. However, a sudden drop in $R(T)$ may indicate the onset of superconductivity, although fluctuating. Additionally, a partial Meissner effect may be detectable once some sections of the sample exhibit phase coherence.

To achieve a large and stable superconducting order parameter, both $|\Psi(\mathbf{r})|^2$ and φ need to be addressed: an ideal superconductor would possess a large $|\Psi(\mathbf{r})|^2$ (linked to the energy gap Δ) and a highly stable φ. From Eqs. 2.10 and 2.16, we know that Δ is inversely proportional to ξ. However, inhomogeneities develop more easily in superconductors with short ξ since it costs less energy to create phase fluctuations. It is therefore interesting to study the stabilization of the phase of the superconducting order parameter in inhomogeneous systems with short ξ, such as low-dimensional materials.

2.2 Superconductivity in Low Dimensions

Superconductivity can take place in all dimensions, however its stability will depend on the interplay between the dimensionality and ξ. If one of the dimensions is reduced to a scale of the order of ξ, the system might not achieve long-range phase coherence.

In a purely one-dimensional (1D) system, the superconducting state is dominated by phase fluctuations, leading to a non-zero resistance. The aim of this section is to study 1D phase fluctuations, discuss how they take place and present earlier works where they have been observed. The concept of quasi-1D (q1D) systems will then be introduced through systems composed of coupled nanowires. Attention will be paid on how such a geometry may be useful to overcome phase fluctuations [8, 9].

2.2.1 One-Dimensional Phase Fluctuations

The phenomenon of phase fluctuations was addressed by Little in 1967 [10], where he studied whether a supercurrent could decay in a ring geometry due to thermodynamic fluctuations. The same year, Langer and Ambegaokar [11] derived a GL free energy barrier at which a transition from one local free energy minimum to another could take place in a thin wire under current-bias, provoking a decay in the supercurrent. These studies showed that a 1D superconducting nanowire at $T < T_c$ exhibits a finite resistance. This phenomenon was explained by analyzing the phase fluctuations of the order parameter (Fig. 2.5). The two local free energy minima differ by a phase of 2π along the length of a wire. Therefore, if a transition from one minimum to another takes place, the phase will slip by 2π, generating a topological defect. This creates a voltage pulse and therefore, a non-zero resistance. The amplitude of the order parameter will be locally suppressed, as $|\Psi|^2 = n_s$.

For $T > 0$, the superconducting order parameter is always subject to both quantum and thermal fluctuations [13]. However, in a bulk material, these fluctuations do not affect the long-range phase coherence. In nanowires with diameter $d \leq \xi$ the order parameter only depends on the position x along the wire. Consequently, even

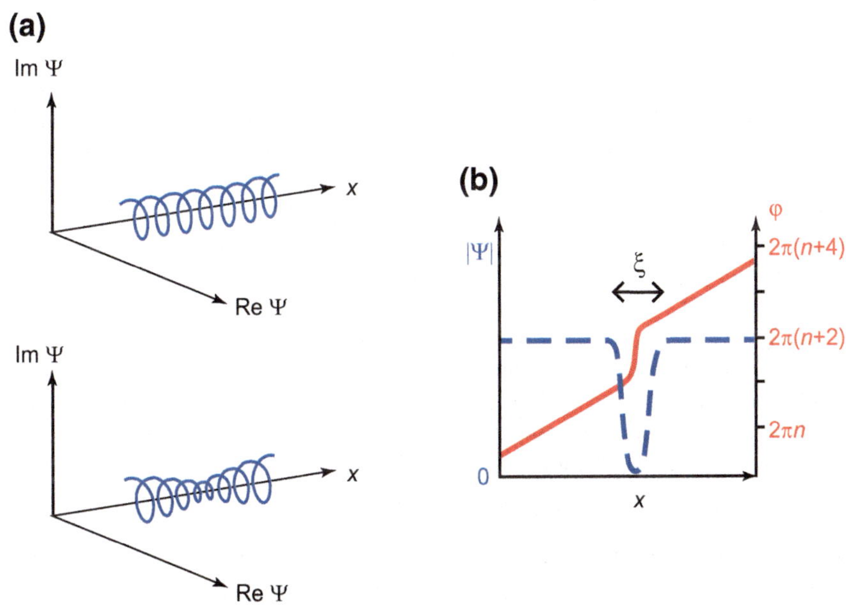

Fig. 2.5 **a** The superconducting order parameter Ψ along a current-carrying wire (x-axis) where the magnitude is perpendicular to the wire length. Top: at equilibrium. Bottom: near the phase slip event. **b** Spatial variation of the phase φ (full line) and the amplitude $|\Psi|$ (dashed line) of the order parameter during the slippage event. The size of a phase slips typically scales with the coherence length ξ. Adapted from Arutyunov et al. [12]

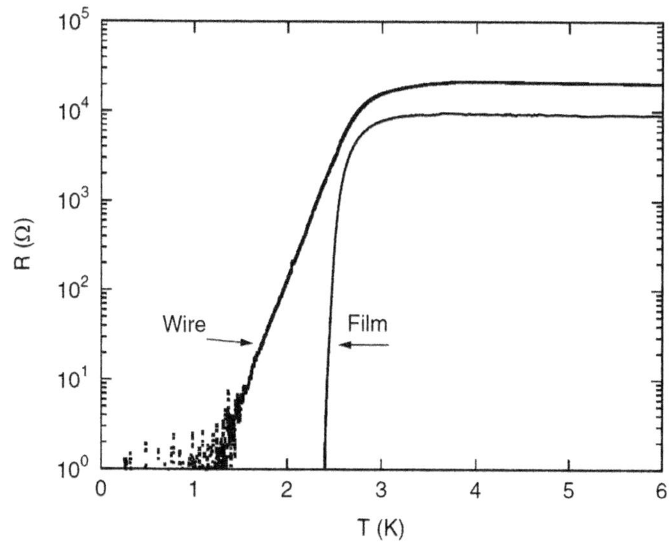

Fig. 2.6 $R(T)$ in a wire (width = 100 nm) and a film (width = 500 μm) composed of amorphous indium-oxide. The wire was prepared by deposition of amorphous indium-oxide on a WS_2 nanotube. $T_c^{wire} = T_c^{film} = 2.8$ K. Data from Johansson et al. [15]

small fluctuations lead to phase slips due to the low energy cost and destroy phase coherence.

These phase slips can be generated either from thermal fluctuations at sufficiently high T, i.e. $T \leq T_c$, called thermally activated phase slips (TAPS), or from quantum fluctuations arising at $k_B T < \Delta_0$ and persisting down to $T = 0$, known as quantum phase slips (QPS).

The effect of reduced dimensionality on the superconducting state has been widely studied from a fundamental point of view and for practical applications. Progress made before 2008 on 1D superconducting transitions are reviewed in Arutyunov et al. [12]. The main observation is a broadening of the superconducting transition close to the 1D limit, as predicted and observed by Giordano [14]. A comparison between a wire near the 1D limit ($d \sim \xi$) and a 2D thin film is presented in Fig. 2.6 [15]. This data set shows how the transition in the 1D wire is widened due to QPS and TAPS contributions.

Thermally Activated Phase Slips

The theory describing TAPS was originally developed by J. Langer, V. Ambegaokar, D. McCumber and B. Halperin [11, 16]. The TAPS contribution to the total resistance can be expressed as:

$$R_{TAPS}(T) = \frac{\pi \hbar^2 \Omega}{2e^2 k_B T} \exp\left(\frac{-\Delta F}{k_B T}\right) \tag{2.18}$$

where ΔF is the energy barrier over which the phase slips are thermally activated.

The time scale of the fluctuations is fixed using a prefactor Ω related to the attempt frequency of random excursions in the superconducting order parameter:

$$\Omega = \frac{L}{\xi}\left(\frac{\Delta F}{k_B T}\right)^{1/2}\frac{1}{\tau_{GL}} \tag{2.19}$$

where L is the length of the wire and $\tau_{GL} = [\pi\hbar/8k_B(T_c - T)]$ is the GL relaxation time.

To obtain an expression of the energy barrier ΔF, Lau et al. [17] developed the following expression for the analysis of the superconducting transition in MoGe nanowires:

$$\Delta F(T) = Ck_B T_c\left(1 - \frac{T}{T_c}\right)^{3/2} \tag{2.20}$$

where C is a dimensionless parameter relating the energy barrier for phase slips ΔF to the thermal energy near T_c:

$$C \approx 0.83\left(\frac{L}{\xi(0)}\right)\left(\frac{R_q}{R_n}\right) \tag{2.21}$$

where $R_q = h/4e^2 = 6.45\,k\Omega$ is the resistance quantum for Cooper pairs and R_n the normal state resistance of the wire.

The total resistance of the 1D filament is evaluated by including the normal state resistance assumed to be temperature independent for simplicity:

$$R_{fit} = \frac{1}{1/R_n + 1/R_{TAPS}} \tag{2.22}$$

Several experiments reported the observation of TAPS in nanowires. For example, Rogachev et al. [18] showed good fits of the TAPS model to Nb nanowires under applied magnetic fields (Fig. 2.7). The peaks visible in the TAPS fits at the highest temperatures are mathematical artifacts, which originate from the fitting equations and can be ignored.

Quantum Phase Slips

The theory describing QPS was mainly developed by A.D. Zaikin et al. [19–21]. QPS arise due to quantum fluctuations in the order parameter and persist even at $T = 0$. The QPS contribution to the resistivity becomes relevant when $k_B T < \Delta_0$ [13]. In a weakly-coupled superconductor ($2\Delta_0 = 3.5k_B T_c$), this corresponds to the temperature range $T < 0.86T_c$.

The QPS contribution to the resistivity may be modeled as [12]:

$$R_{QPS} = A_Q B_Q \frac{R_q^2}{R_n}\frac{L^2}{\xi(0)^2}\exp\left[-A_Q \frac{R_q}{R_n}\frac{L}{\xi}\right] \tag{2.23}$$

where A_Q and B_Q are constants.

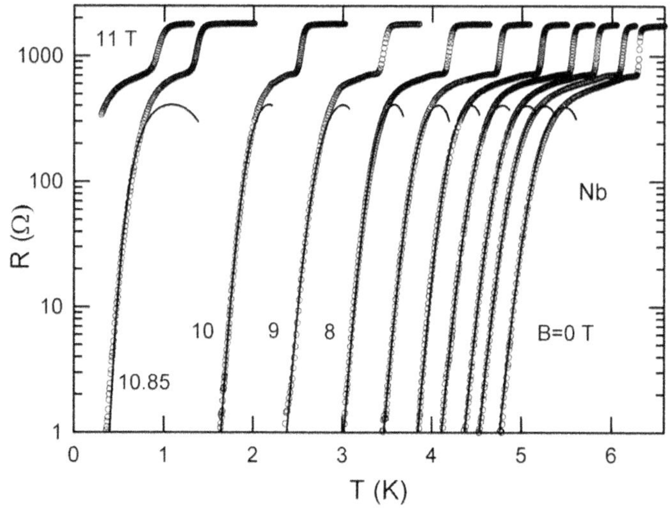

Fig. 2.7 $R(T)$ in Nb nanowires of thickness 8 nm, length 120 nm, and with $R_n = 700\ \Omega$. The superconducting transition are fitted using TAPS model (solid lines) for different fixed magnetic fields. Data and fits from Rogachev et al. [18]

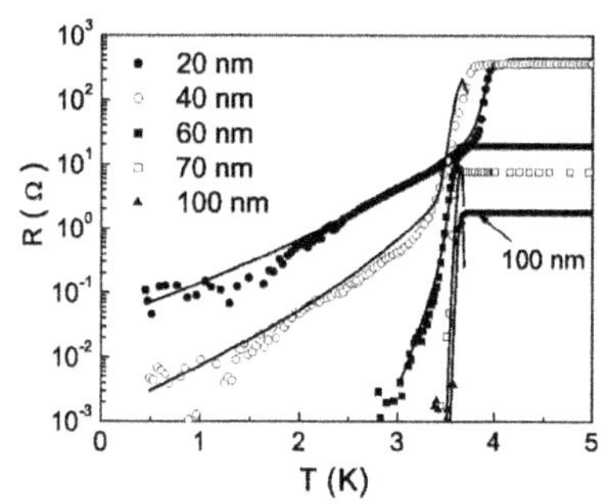

Fig. 2.8 $R(T)$ in Sn nanowires with diameter between 20 and 100 nm, ($\xi_0 \sim 200$ nm). The 70–100 nm wires data were fitted with TAPS model. The 20, 40 and 60 nm wires data were fitted with TAPS+QPS model. Fits are shown as solid lines. Data and fits from Tian et al. [22]

Finally, the total resistance of a 1D wire is evaluated as the sum of the quasiparticle contribution and the TAPS + QPS terms:

$$R_{fit} = \frac{1}{1/R_n + 1/\left(R_{TAPS} + R_{QPS}\right)} \tag{2.24}$$

The effect of QPS on transport measurements is shown in Fig. 2.8, where thinner wires exhibit combined QPS and TAPS behavior. The TAPS model is used near T_c while the QPS model applies at lower temperature. The visible tail in the resistance at lower temperature is a typical signature of QPS [12].

2.2.2 *Quasi-One-Dimensional Superconductors*

Experiments on pure 1D nanowires provided considerable information on the stability of the superconducting state as a function of dimensionality. However, 1D superconductors are not suitable for applications due to fluctuations and size limitations (as $d \ll \xi$, where in general $\xi < 1$ μm). It would still be of great interest to develop superconducting materials with a 1D morphology for applications such as cables, but with a high phase stiffness. Also, low-dimensional systems with short coherence length exhibit attractive properties such as high critical fields (Eq. 2.17).

In this light, it is appealing to join several 1D superconducting wires and allow them to electrically couple, combining both attractively short ξ and phase-coherent superconductivity.

While the nature of the coupling will be discussed in the next section, one can already visualize the structure of a q1D system as presented in Fig. 2.9. Q1D systems can be considered as an array of filaments, where the electron dynamics within a chain are characterized by energy scales much larger than the coupling between the chains.

The electron dynamics are defined by the electron hopping energies, which describe the overlap of the electron wavefunctions. A q1D material is expected to exhibit large transport along the wires (i.e. longitudinal hopping energy $t_{||}$) compared to between the wires (i.e. transverse hopping energy $t_{\perp}$). While $t_{||}$ and $t_{\perp}$ can be calculated from band structure using density functional theory, it is also possible to obtain a good estimate of the anisotropy of the electronic transport from the ratio of the longitudinal and transverse conductivity $\sigma_{||}/\sigma_{\perp}$.

In such systems, electrons can hop from one wire to another under favorable conditions: coherent electron hopping can only occur if $t_{\perp}$ is sufficiently large.

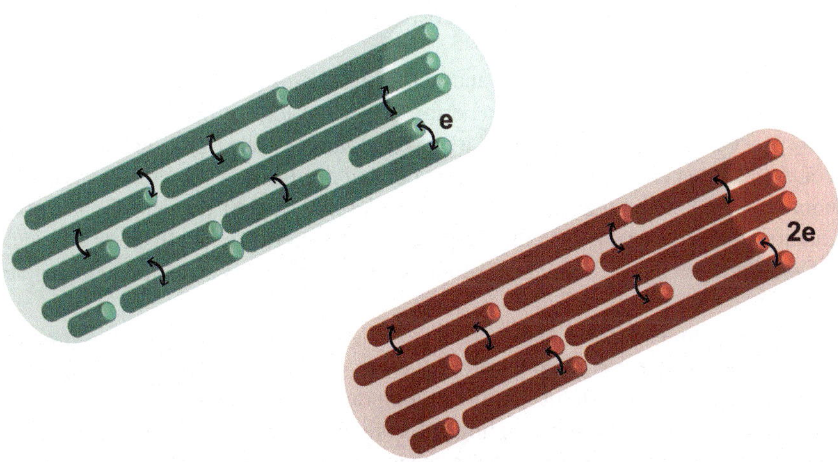

Fig. 2.9 Schematic of a q1D system, where electrons in the normal state (e, left) and Cooper pairs in the superconducting state (2e, right) can hop from one wire to another. The electronic properties of this system depend whether electrons or Cooper pairs can hop coherently between the 1D wires

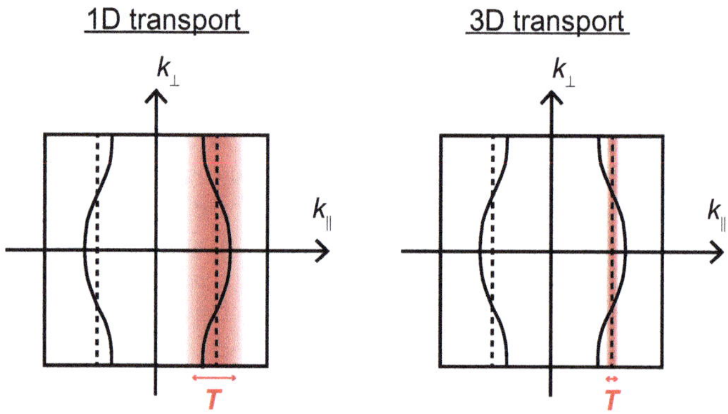

Fig. 2.10 Schematic of an open Fermi surface in a q1D system where $t_\perp \ll t_\parallel$. **Left**: if the external energy (here temperature T) is larger than the warping of the Fermi surface $t_\perp$, the system cannot feel the warping and the electrons evolve in 1D. **Right**: at low external energy (here $T < t_\perp$), the system feels the extra dimensions of the dispersion, and the electrons evolve in 3D. Adapted from Giamarchi [23]

The value of $t_\perp$ corresponds to the warping of the Fermi surface, where a smaller $t_\perp$ equals a flatter Fermi surface. From here, one may introduce the concept of a dimensional crossover as a function of the system's Fermi surface anisotropy and the variation of external energies, such as temperature. Understanding this dimensional crossover is the key to the physics of q1D systems.

Let us first consider a system with negligible electron interactions. In Fourier space, the kinetic energy is expressed as [23]:

$$\varepsilon(\mathbf{k}_\parallel, \mathbf{k}_\perp) = -t_\parallel \cdot \cos(\mathbf{k}_\parallel a) - t_\perp \cdot \cos(\mathbf{k}_\perp b) \tag{2.25}$$

where a, b are measured parallel and perpendicular to the chains, respectively. If $t_\perp \ll t_\parallel$, Eq. 2.25 expresses an open Fermi surface. As long as the external excitation energy (such as temperature) is of a larger scale than the warping of the Fermi surface (controlled by $t_\perp$), the warping is washed out and the electron wavefunctions evolve uniquely in 1D, along the chains (1D transport, Fig. 2.10). At $T < t_\perp$ all correlation functions become sensitive to the warping (3D transport, Fig. 2.10). Hence in ideal q1D systems, a 1D to 3D electronic dimensional crossover is expected to take place at $T = t_\perp$.

The discussion on dimensional crossover may now be extended to the superconducting state. If the wires are superconducting, Cooper pairs may hop coherently between the chains depending on the amplitude of the tunneling energy, which is determined by the Josephson coupling perpendicular to the wires. Josephson tunneling is a second order process which may occur only for $T < t_\perp^2/t_\parallel$ [23]: now bosonic pairs are considered instead of fermions. For most q1D materials, the superconducting transition takes place at $T_c < t_\perp^2/t_\parallel$ and the transition is 3D: when the Cooper pairs are formed, they can hop from one chain to the other. In this case, the transition

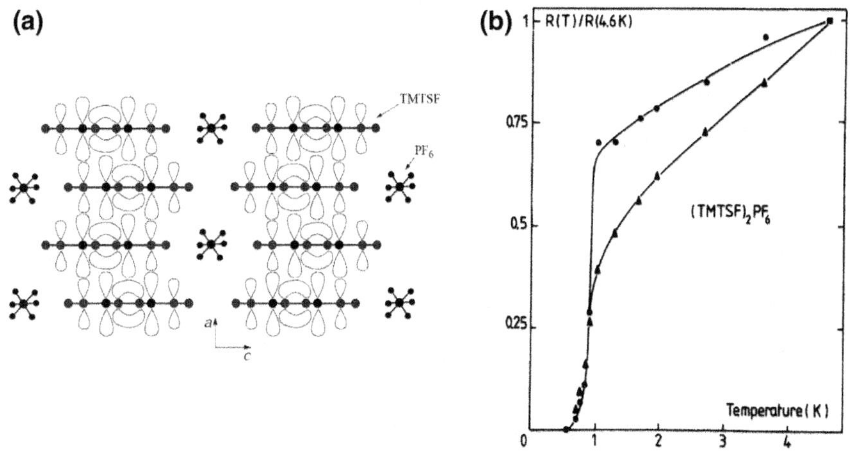

Fig. 2.11 Superconductivity in the q1D Bechgaard salt $(TMTSF)_2PF_6$. **a** Side view of the crystal structure of $(TMTSF)_2PF_6$. Each TMTSF molecule is shown with the electron orbitals. The material is the most conductive along the a axis. From Mourachkine [7]. **b** Normalized resistance $R(T)/R(4.6\,\mathrm{K})$ in two different samples with an applied pressure of 12 kBar. From Jérome et al. [24]

is identical to that of a bulk superconductor, even though the transport remains highly anisotropic.

On the other hand, if the anisotropy and the pairing interaction are both sufficiently large, the onset temperature for superconducting fluctuations T_{ons} may be greater than $t_{\perp}^2/t_{\parallel}$. In this case, a 2-step superconducting transition occurs [8, 9]. First at T_{ons}, 1D superconductivity develops along the wires and is dominated by fluctuations due to phase slips. Then, at $T \sim t_{\perp}^2/t_{\parallel}$, Cooper pairs hop between the chains, suppressing phase slips and establishing phase coherence. It is only upon the establishment of transverse coupling that the state can be truly coherent as otherwise, the phase slips of the order parameter generate non-zero voltage pulses at some points in the wires.

A one step superconducting transition has been reported in charge transfer organic superconductors known as the Bechgaard salts [24]. These materials are layered with a strong in-plane anisotropy, making them electronically q1D (Fig. 2.11a). The Bechgaard salt $(TMTSF)_2PF_6$ was shown to superconduct below $T_c = 0.9\,\mathrm{K}$ under a pressure of 12 kBar (Fig. 2.11b) [24].

This material shows a relatively large ratio between the perpendicular and parallel hopping energies [25]: $t_{\parallel} = 3000\,\mathrm{K}$ and $t_{\perp} = 230\,\mathrm{K}$. As the Cooper pair dimensional crossover takes place at $t_{\perp}^2/t_{\parallel} = 18\,\mathrm{K}$, the observed superconducting transition is the one of a 3D material: the superconducting transition drops sharply in contrast to that observed in 1D nanowires.

A two-step superconducting transition has been observed in carbon nanotubes embedded in a zeolite matrix [26] (Fig. 2.12). For this engineered material, no hopping energies have been theoretically studied, but various experimental works report interesting features inherent to q1D superconducting transitions [26].

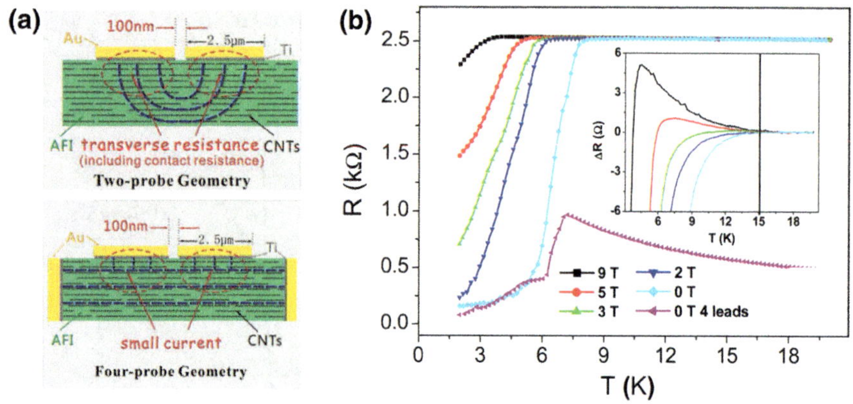

Fig. 2.12 Superconductivity in a q1D matrix of carbon nanotubes. **a** Sketches of a sample where the different components are indicated. Two contact modes were used: two-probe (top) and four-probe (down). The dashed blue lines represent the current paths. **b** $R(T)$ at $I = 1\,\mu A$. The magenta curve is the four-probe data and the others are two-probe data. The inset shows a magnified view of the superconducting transition. From Wang et al. [26]

Transport measurements were performed using two- and four-probe contact configurations (Fig. 2.12a). The suppression of the transition to lower temperature with applied magnetic field indicates its superconducting origin (Fig. 2.12b).

In both types of contact configuration, a hump appearing midway through the transition is observed, although it is clearer in the 4-probe experiment. This is a signature of the 1D to 3D dimensional crossover.

In the analysis of these carbon nanotubes matrices, the crossover is described to take place via a Berezinskii–Kosterlitz–Thouless (BKT) transition, where phase coherence is established via phase locking in the transverse plane perpendicular to the nanotubes. A BKT transition describes a 2D transition from free vortices and antivortices to vortex-antivortex pairs with decreasing temperature (see Chap. 7 for more details).

2.2.3 *Superconducting Transition in a Quasi-One-Dimensional System*

Q1D systems such as filamentary materials cannot be analyzed as single quantum wires with pure phase slips theories since their diameter is expected to be much greater than ξ. However, they are composed of multiple 1D filaments, each dominated by phase slips at $T > t_\perp^2/t_\parallel$. The 1D TAPS model is adapted to describe a q1D network of uncoupled filaments. As superconductivity is expected to be fluctuating only close to T_{ons} (phase coherence develops at $T \sim t_\perp^2/t_\parallel$), QPS are not considered in this analysis.

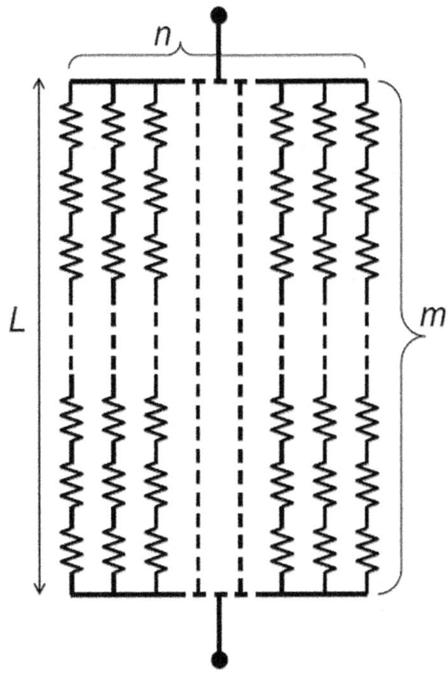

Fig. 2.13 Uniform $m \times n$ array of identical resistors, used to adapt the 1D TAPS theory to a q1D filamentary system of length L

The q1D material is modeled as a $m \times n$ array of 1D filaments, i.e. a $m \times n$ network of identical resistors (representing wires) with resistance R_n (Fig. 2.13). The choice of a 2D array to describe a crystal existing in 3 spatial dimensions is made purely for simplicity, since the resistance of a line of n resistors in parallel is equivalent to that of a $\sqrt{n} \times \sqrt{n}$ lattice. The total resistance of the $m \times n$ array is $R_N = mR_n/n$, where R_N corresponds to the experimentally measured normal state resistance of the q1D material.

Below T_{ons}:

$$\frac{1}{R} = \frac{1}{R_N} + \frac{n}{m}\frac{1}{R_{TAPS}} \tag{2.26}$$

where R_{TAPS} is the TAPS resistance of a single superconducting filament of length L, which appeared in Eq. 2.22. The first step is to re-express the parameter C from Eq. 2.21 in terms of R_N:

$$C \approx 0.83\left(\frac{Lm}{n\xi(0)}\right)\left(\frac{R_q}{R_N}\right) \tag{2.27}$$

Then, an effective length $L_{eff} = Lm/n$ is defined and the phase fluctuation attempt frequency Ω is renormalized accordingly. This yields:

$$\frac{1}{R} = \frac{1}{R_N} + \frac{1}{R_{eff}} \tag{2.28}$$

Here, R_{eff} is the total TAPS contribution to the resistance, controlled by the fitting parameter C.

This shows that the standard TAPS model can describe a q1D geometry using a geometrically renormalized length L_{eff}. Note that in the present model, a single average characteristic length has been chosen for simplification. This is likely to be realistic for macroscopic q1D systems comprising a multitude of such filaments.

To the best of our knowledge, it is the first time such adaptation of the 1D TAPS model is proposed. Its validity and application to the experimental results will be discussed in Chaps. 6 and 7, where this model is used to estimate the average diameter of a filament in a q1D crystalline material.

2.3 Josephson Coupling

Electrons can tunnel through an insulating barrier separating two normal metals, with their tunnel current density decaying exponentially along the barrier. If the normal metal sections are replaced by superconductors, the normal electrons at the Fermi level are replaced by Cooper pairs at $T < T_c$, and tunneling would be expected to take place only under a sufficiently large applied voltage which would break the pairs, i.e. for $eV \geq 2\Delta_0$. However, B. Josephson showed in 1962 that Cooper pairs could tunnel through the barrier with the same probability as single electrons [27].

A simple way to understand this effect is to consider a superconductor - insulator - superconductor (SIS) junction as a molecule of H_2: the molecular binding exists due to the overlap of the electron wavefunctions of the two H atoms. In the same way, the two superconductors will be coupled when the Cooper pairs wavefunctions overlap. The macroscopic tunneling wave function is a coherent state which propagates through the barrier.

Josephson coupling was initially described for SIS junctions, but it can be extended to more general cases [1, 28]: if two superconducting regions are separated by a non-superconducting thin barrier, called a weak link, a coupling can be established via tunneling of Cooper pairs through the weak link (Fig. 2.14). Weak links can be insulating or metallic barriers, narrow constrictions, or grain boundaries in an inhomogeneous material [28].

2.3.1 *The Josephson Junction and Critical Current*

In a Josephson junction, a zero resistance supercurrent I_s can develop between two superconducting regions:

$$I_s = I_c \sin(\Delta\varphi) \tag{2.29}$$

where $\Delta\varphi$ is the phase difference and I_c the critical current corresponding to the maximum supercurrent the junction can withstand.

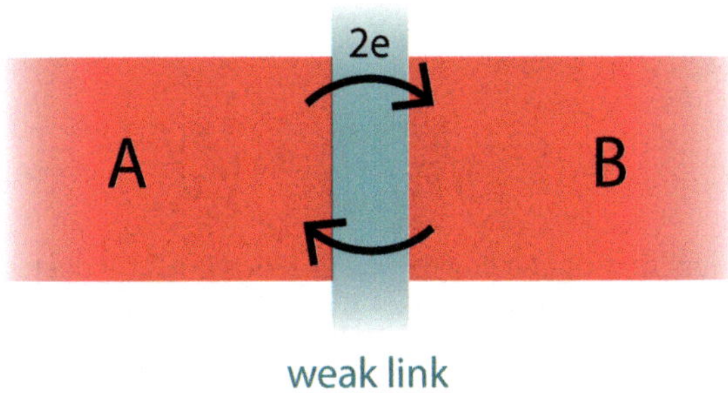

Fig. 2.14 A Josephson junction: two weakly-coupled superconductors A and B (red) are separated by a weak link (blue). Phase-coherent superconductivity is established by Cooper pairs (2e) tunneling through the weak link

If such a junction possesses a voltage difference V, $\Delta\varphi$ evolves as:

$$d\left(\Delta\varphi\right)/dt = 2eV/\hbar \tag{2.30}$$

leading to an alternating I_s with amplitude I_c and frequency $2eV/h$.

Conventionally, the case when $V = 0$ is known as the dc Josephson effect, while $V \neq 0$ is the ac Josephson effect [1].

The value of the critical current I_c defines the strength of the phase coupling. $I_c(T)$ in a Josephson junction was derived by V. Ambegaokar and A. Baratoff [29]:

$$I_c(T)R_N = \frac{\pi\Delta(T)}{2e}\tanh\left(\frac{\Delta(T)}{2k_BT}\right) \tag{2.31}$$

where R_N is the normal state resistance.

This equation is an important general result implying that at $T = 0$, $I_cR_N = \frac{\pi\Delta(0)}{2e}$: in an ideal Josephson junction, the critical current (and hence the coupling) is maximized at $T = 0$ [1].

In a homogeneous junction under an applied magnetic field H, oscillations in I_c will be observed with a period of $\pi Hwt/\Phi_0$ where w is the width and t the thickness of the junction [30]. These oscillations are usually referred to as a Fraunhofer pattern [1, 31] and characterize the homogeneity of the junction. Consequently, if a material is composed of inhomogeneous Josephson junctions of varying dimensions, a Fraunhofer pattern will not be visible in the experimental data.

2.3.2 *The Josephson Energy*

The potential energy accumulated in a junction is defined as the coupling energy, or Josephson energy E_J [1]:

$$E_J \equiv \frac{\hbar I_c}{2e} = \frac{\Phi_0 I_c}{2\pi} \tag{2.32}$$

E_J characterizes the strength of the coupling. By combining Eqs. 2.31 and 2.32, the evolution of $E_J(T)$ is given by:

$$E_J(T) = \frac{\hbar}{2eR_N}\Delta(T)\tanh\left(\frac{\Delta(T)}{2kT}\right) \tag{2.33}$$

A Josephson junction can be considered as a non-linear inductance which accumulates energy from the flowing current. E_J represents the accumulated energy, in contrast to a real inductance where the current generates a magnetic field.

2.3.3 *Internal Josephson Junctions in Low-Dimensional Superconductors*

Josephson coupling in q1D systems will take place only for $T \sim t_{\perp}^2/t_{\parallel}$ and a direct relationship can be established between the hopping energies and the Josephson energy.

Let us consider a Josephson junction composed of two superconducting wires, where the surrounding matrix constitutes the weak link. Here, E_J is proportional to the spatial overlap of the Cooper pair wavefunctions from each wire which is controlled by their hopping energies. The minimum energy required to establish Josephson coupling in such junctions is:

$$E_J^{min} = \frac{\hbar}{2eR_N}\Delta(T_J)\tanh\left(\frac{\Delta(T_J)}{2kT_J}\right) \tag{2.34}$$

where $T_J \sim t_{\perp}^2/t_{\parallel}$ is the Josephson temperature, corresponding to the maximum temperature at which Josephson coupling can take place in the system.

A real q1D material is composed of a multitude of such two-wire Josephson junctions, where each junction is characterized by a singular R_N and accordingly E_J^{min}. Additionally, $t_{\perp}^2/t_{\parallel}$ defines theoretically the minimum temperature at which the coupling takes place: any e-e correlation or renormalization of $t_{\perp}$ or $t_{\parallel}$ will modify the real value of T_J. This is therefore only a very approximate approach which however, provides a good theoretical ground to analyze real q1D systems.

2.4 Disorder and Superconductivity

Disorder such as the introduction of impurities or atomic vacancies is present in most functional superconductors. Numerous theoretical and experimental studies on the effect of disorder on superconductivity have been performed since the 1950s [32–48], leading to either suppression, persistence, or enhancement of superconductivity.

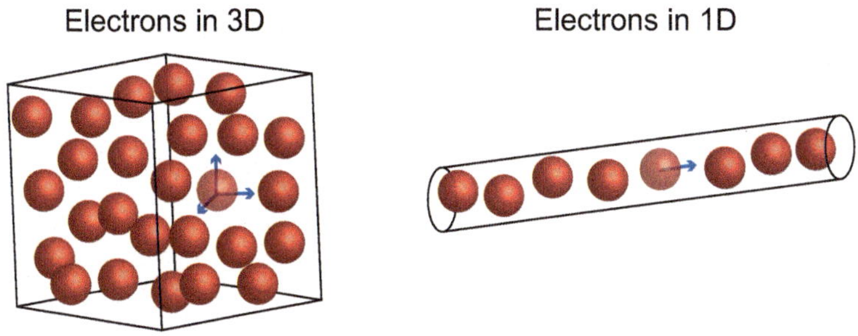

Fig. 2.15 Impact of reduced dimensions on electron transport. As the dimensionality decreases, electrons confinement increases

The presence of disorder in a conductive system provokes electron-impurity scattering, changing the ballistic motion of free electrons to diffusive and reducing the electron mean-free path l. If $\xi_0 \ll l$, a superconductor is considered to be in the clean limit. On contrary, if $k_F \ll l \ll \xi_0$, the material is considered a dirty superconductor (where k_F is the Fermi wave vector). In the dirty limit, the effective coherence length is reduced as $\xi \approx \sqrt{\xi_0 l}$ [49].

When l is reduced by disorder below a critical value $l_c \sim 1/k_F$, the diffusive motion of electrons is suppressed and extended Bloch waves become localized. This was first proposed by Anderson [32] and is known as Anderson localization or strong localization.

In a system with reduced dimensions, the effect of disorder will be greatly enhanced. Electron path possibilities decrease with decreasing dimensions (Fig. 2.15) and even a small level of disorder in 1D will affect electron transport.

In low-dimensional systems, small levels of disorder are expected to induce Anderson localization [50]. Therefore, weakly disordered low-dimensional systems can be considered as effectively strongly disordered.

In the presence of strong disorder, electrons near the Fermi level spatially localize, creating an insulating ground state at $T = 0$. However, electrons near the Fermi level in the presence of even the smallest net attraction (created by phonon interactions) can form Cooper pairs, leading to a superconducting ground state.

Hence, a disordered metal with a superconducting instability experiences competition between the superconducting and insulating ground states as $T \to 0$.

2.4.1 *Electron Localization in Three Dimensions*

In a bulk material with weak disorder, electrons are normally scattered but remain extended through the entire space. When $l \sim l_c$, electron wavefunctions become localized and their amplitude drops exponentially with distance from the localization center $\mathbf{r}_0$ [49]:

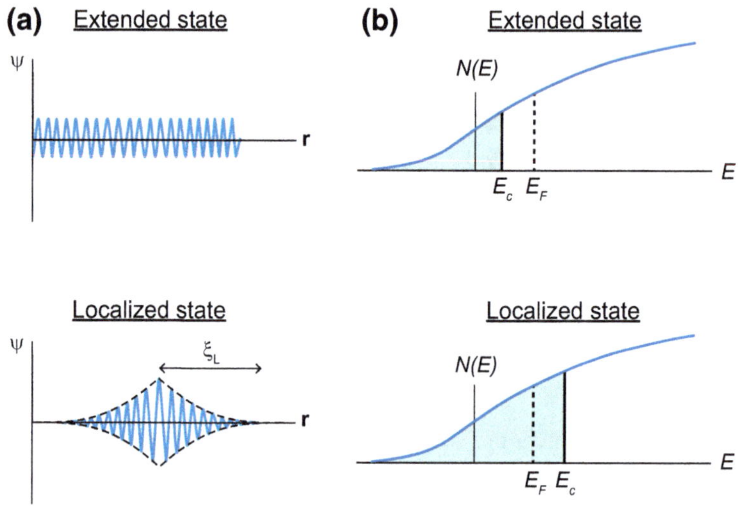

Fig. 2.16 Electron localization in a disordered system. **a** The extended electron wavefunction ψ becomes localized in the presence of large disorder. The localization length ξ_L corresponds to the exponential decay of the wavefunction along **r**. **b** Electron density of states $N(E)$ near the band edge, where the localized states are separated from the extended states by the mobility edge E_c. If the Fermi energy $E_F > E_c$, the system is metallic. Ff $E_F < E_c$ the system is insulating and transport can only take place via electron hopping between localized states. Adapted from Sadovskii [49]

$$|\psi| \sim \exp\left(\frac{|\mathbf{r} - \mathbf{r}_0|}{\xi_L}\right) \tag{2.35}$$

The exponential drop can be visualized as an envelope surrounding the wavefunction with a characteristic decay length equal to the localization length ξ_L. The difference between an extended and a localized wavefunction is depicted in Fig. 2.16a. The stronger the disorder, the shorter is ξ_L.

In a disordered material subject to Anderson localization, a critical energy E_c, known as the mobility edge, separates the extended from the localized states (Fig. 2.16b). For low levels of disorder, $E_F > E_c$ and the wavefunctions remain extended. For large levels of disorder, the Fermi energy $E_F < E_c$ and the system is localized. When $E_F = E_c$ the system undergoes the so-called Anderson transition from extended to localized states, which has been experimentally demonstrated to be continuous: no sudden drop of the conductivity appears in the transport measurement. This continuity is achieved due to thermally-activated electron hopping between localized states. Therefore, the system is fully insulating (with no transport enabled) at $T = 0$.

The separation δ between the energies of two localized electron wavefunctions can be estimated as [49]:

$$\delta \approx \frac{1}{N(E_F)\xi_L} \tag{2.36}$$

This expression provides an energy scale for determining whether a disordered material subject to localization can still undergo a superconducting transition: if $\Delta_0 >> \delta$, a superconducting ground state is favorable to an insulating one. Superconductivity can therefore take place within localized states assuming $\xi_0 \leq \xi_L$.

2.4.2 Anderson Theorem

Theoretical studies on the interplay between disorder and superconductivity were first performed by A.A. Abrikosov and L.P. Gork'ov [33] and P.W. Anderson [34] in 1959. They showed that the transition from free electron motion to diffusive motion does not affect the superconducting state, as long as the impurities were non-magnetic, i.e. as long as disorder does not break time-reversal invariance. This is nowadays known as the Anderson theorem. This theorem was derived for the case of conventional superconductors: BCS-like with an isotropic energy gap (s-wave superconductors).

In conventional superconductors, it is expected that the value of the superconducting critical temperature T_c does not suffer from the reduction of l provoked by electron scattering. Indeed, the BCS expression of T_c only depends on the Debye frequency and the electron-phonon coupling (Eq. 2.5). However, this argument stays valid only for $l > l_c$, i.e. when electrons do not experience localization. When the electron wavefunctions become localized, other effects have to be taken into account such as electron hopping or Coulomb repulsion.

2.4.3 Metal-Insulator Transition in Two Dimensions

At sufficiently strong disorder, the Anderson theorem does not apply anymore: either electron localization or the combination of electron scattering and Coulomb repulsion leads to the suppression of T_c. The explanation for the latter effect was provided in 1982 by S. Maekawa et al. [35, 36], after several superconductor-insulator transitions were observed in 2D disordered materials.

The presence of large disorder can lower T_c in thin films as:

$$\ln \frac{T_c}{T_{c0}} \approx -\frac{g_1}{3} R \ln^3 \left(\frac{\omega_D}{T_c} \right) \tag{2.37}$$

where $g_1 = N(0)V$ is the dimensionless constant of the local interactions V (such as Coulomb repulsion and Cooper pair attraction) and R the surface resistance of the film. In the presence of Coulomb interaction, $g_1 = 1$ and Eq. 2.37 describes the suppression of T_c by disorder (with increasing R).

This effect was studied by A. Finkelstein [37] who showed the suppression of superconductivity in amorphous MoGe films (Fig. 2.17). Superconductor-to-insulator transitions in disordered systems are widely caused by dimensionally-enhanced Coulomb pair breaking due to electron scattering. Here, disorder directly

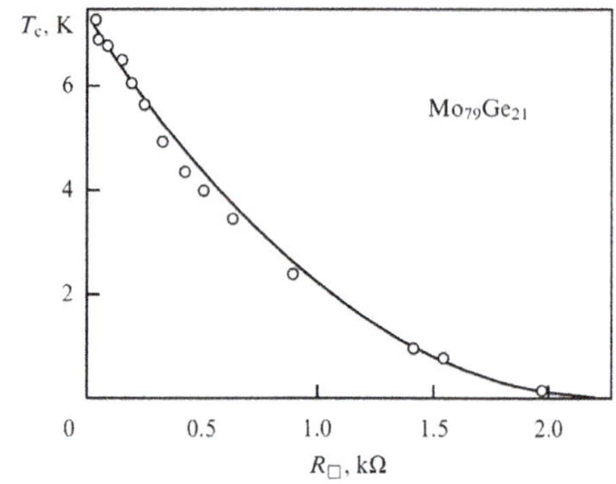

Fig. 2.17 Suppression of the superconducting transition temperature T_c with increasing surface resistance R in amorphous 2D $Mo_{79}Ge_{21}$ films. The fit (solid line) is derived from Eq. 2.37. From Finkel'stein [37]

controls the sheet resistance of the amorphous film, which is inversely proportional to its thickness. The thickness of the film controls the anomalous diffusion present in disordered metals: the thinner the film, the smaller the screening of the renormalized Coulomb interaction and the larger the impurity scattering. This combined effect leads to a suppression of T_c.

Coulomb pair-breaking is not the only mechanism which can destroy superconductivity in 2D thin films. Emergent granularity, i.e. macroscopic inhomogeneity in a thin film, also tends to suppress the superconducting state. A granular material consists of grains of a characteristic size b embedded in a matrix of normal metal or insulator with a certain resistance (Fig. 2.18). A granular material can develop a superconducting state if the average energy spacing δ_ϵ between the electron wave-functions inside a grain is smaller than the superconducting gap Δ_0:

$$\delta_\epsilon = \frac{1}{N(0)b^3} < \Delta_0 \tag{2.38}$$

This equation therefore defines the minimum size of a grain for a granular material to develop superconductivity.

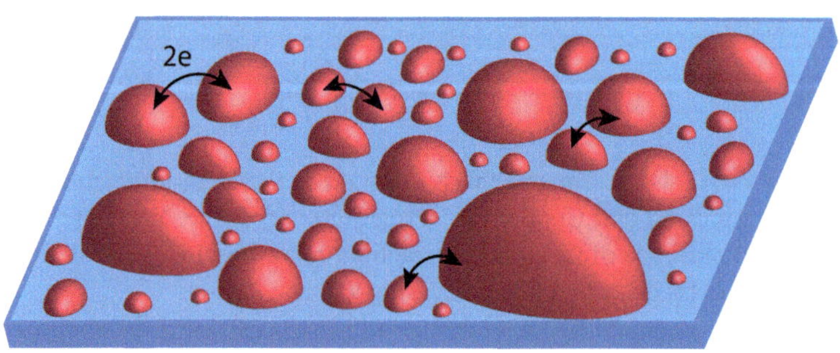

Fig. 2.18 Schematic of a granular material, where the grains need to have a minimum size to develop a superconducting state (Eq. 2.38). Coherent superconductivity can be established by tunneling of Cooper pairs (2e) between the grains

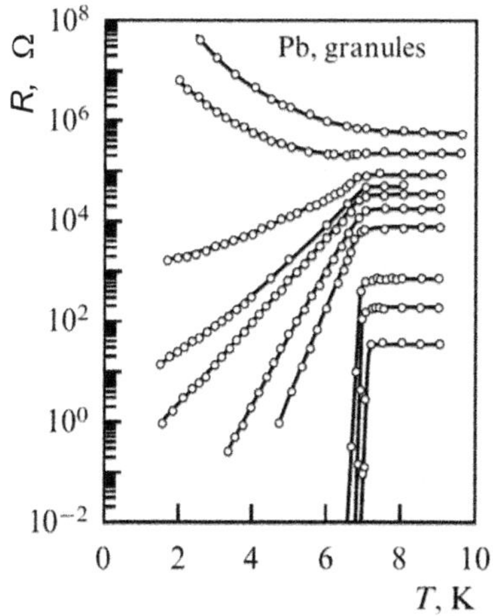

Fig. 2.19 Suppression of the superconducting transition T_c by disorder in 2D granular Pb films deposited on ultra-cold SiO substrate. $R(T)$ decreases upon increasing the film thickness (from top to bottom). From Frydman [39]

A coherent superconducting state will be established when Cooper pairs can tunnel between the grains. This is controlled by the distance between the grains and the resistance of the matrix material. Granular superconductivity has been experimentally investigated in various systems, such as lead films deposited on ultra-cold SiO substrates [39]. Transport results from this material are presented in Fig. 2.19.

This growth technique produces island-like films, where each island has an average size b directly controlled by the thickness of the film. With increasing film thickness (i.e. decreasing R), an insulator-to-superconductor quantum phase transition occurs. Below a critical thickness the films are insulating, while above this critical size all the films show $T_c \approx 7\,\text{K}$. This means that the superconducting transition inside each grain takes place at the same temperature. While the intermediate curves show the emergence of fluctuating superconductivity (with a non-zero resistive state), phase-coherent superconductivity develops in the three thickest films (three bottom curves). This is a signature of an established tunneling between the grains [49].

In addition to the amorphous Mo-Ge films and granular Pb films presented in Figs. 2.17 and 2.19, superconductor-insulator transitions have been reported in Bi films [40], ultrathin Be films [41], thin TiN films [42], and amorphous thick In-O films [43] to cite only a few examples.

2.4.4 New Approaches and Theories

Considering again the equation developed by Maekawa (Eq. 2.37), the evolution of T_c can be studied in the absence of Coulomb interactions. When the Coulomb repulsion is strongly suppressed, g_1 describes the attraction between electrons in Cooper pairs and takes the value $g_1 < 0$, which is defined by the BCS theory as $g_1^{-1} = \ln(\omega_D/T_c)$, leading to [47]:

$$\ln \frac{T_c}{T_{c0}} \approx \frac{1}{3} R \ln^2 \frac{\omega_D}{T_c} \tag{2.39}$$

This expression indicates a possible enhancement of T_c with R in a disordered superconductor if the Coulomb repulsion is negligible or suppressed.

Such an enhancement has recently been theoretically predicted to develop in systems where disorder induces Anderson localization [44–48]. Close to the mobility edge, in the critical region between extended and localized states, the electron wavefunctions are subject to multifractality. In order to still extend throughout the system while being partially localized, each wavefunction occupies a fractal dimension in real space, developing a scale-invariant pattern such as the one observed in a snowflake or a fern leaf.

The combination of a suppressed Coulomb repulsion and the presence of Anderson localization (hence multifractality of the electron wavefunctions) in a disordered superconductor has been predicted to enhance e-e interactions and, therefore, the superconducting pairing energy.

Until now, this effect has only been theoretically predicted and no experimental data has confirmed this hypothesis. The work presented in Chap. 6 and published in Petrović et al. [9] is the first experimental study compatible with the multifractal theories. This suggests that q1D systems could be suitable candidates to observe such an enhancement of superconductivity.

References

1. M. Tinkham, *Introduction to Superconductivity*. 2nd edn. (Dover Publications, 1996)
2. F. London, H. London, The electromagnetic equations of the supraconductor. Proc. R. Soc. Math. Phys. Eng. Sci. **149**(866), 71–88 (1935)
3. F. London, *Superfluids* (Wiley, New York, 1950)
4. J. Bardeen, L.N. Cooper, J.R. Schrieffer, Theory of superconductivity. Phys. Rev. **108**(5), 1175–1204 (1957)
5. L.D. Landau, D. Ter-Haar, *Collected Papers of L. D. Landau* (Pergamon Press, Oxford, 1965)
6. A.A. Abrikosov, Magnetic properties of superconductors of the second group. Sov. Phys. JETP **5**(6) (1957)
7. A. Mourachkine, *Room-Temperature Superconductivity* (Cambridge International Science Publishing, Cambridge, 2006)
8. D. Ansermet, A.P. Petrović, S. He, D. Chernyshov, M. Hoesch, D. Salloum, P. Gougeon, M. Potel, L. Boeri, O.K. Andersen, C. Panagopoulos, Reentrant phase coherence in superconducting nanowire composites. ACS Nano **10**(1), 515–523 (2016)
9. A.P. Petrović, D. Ansermet, D. Chernyshov, M. Hoesch, D. Salloum, P. Gougeon, M. Potel, L. Boeri, C. Panagopoulos, A disorder-enhanced quasi-one-dimensional superconductor. Nat. Commun. **7**, 12262 (2016)
10. W.A. Little, Decay of persistent currents in small superconductors. Phys. Rev. **156**(2), 396–403 (1967)
11. J.S. Langer, V. Ambegaokar, Intrinsic resistive transition in narrow superconducting channels. Phys. Rev. **164**(2), 498–509 (1967)
12. K.Y. Arutyunov, D.S. Golubev, A.D. Zaikin, Superconductivity in one dimension. Phys. Rep. **464**(1–2), 1–70 (2008)

13. F. Altomare, A. Chang, *One-Dimensional Superconductivity in Nanowires*, (Wiley, 2013)
14. N. Giordano, Dissipation in a one-dimensional superconductor: evidence for macroscopic quantum tunneling. Phys. Rev. B **41**(10), 6350–6365 (1990)
15. A. Johansson, G. Sambandamurthy, D. Shahar, N. Jacobson, R. Tenne, Nanowire acting as a superconducting quantum interference device. Phys. Rev. Lett. **95**(11), 1–4 (2005)
16. D.E. McCumber, B.I. Halperin, Time scale of intrinsic resistive fluctuations in thin superconducting wires. Phys. Rev. B **1**(3), 149–157 (1970)
17. C.N. Lau, N. Markovic, M. Bockrath, A. Bezryadin, M. Tinkham, Quantum phase slips in superconducting nanowires. Phys. Rev. Lett. **87**(21), 217003 (2001)
18. A. Rogachev, A. Bollinger, A. Bezryadin, Influence of high magnetic fields on the superconducting transition of one-dimensional Nb and MoGe nanowires. Phys. Rev. Lett. **94**(1), 017004 (2005)
19. A.D. Zaikin, D.S. Golubev, A. van Otterlo, Quantum phase slips and transport in ultra-thin superconducting wires. Phys. Rev. Lett. **78**(8), 1552 (1997)
20. A.D. Zaikin, D.S. Golubev, A. van Otterlo, G.T. Zimanyi, Quantum fluctuations and dissipation in thin superconducting wires. Phys. Uspekhi **41**(2), 226–230 (1998)
21. D.S. Golubev, A.D. Zaikin, Quantum tunneling of the order parameter in superconducting nanowires. Phys. Rev. B **64**(1), 014504 (2001)
22. M. Tian, J. Wang, J. Kurtz, Y. Liu, M. Chan, T. Mayer, T. Mallouk, Dissipation in quasi-one-dimensional superconducting single-crystal Sn nanowires. Phys. Rev. B **71**(10), 104521 (2005)
23. T. Giamarchi, Theoretical framework for quasi-one dimensional systems. Chem. Rev. **104**(11), 5037–5056 (2004)
24. D. Jerome, A. Mazaud, M. Ribault, K. Bechgaard, Superconductivity in a synthetic organic conductor (TMTSF)2PF6. J. de Phys. Lett. **41**(4), 95–98 (1980)
25. J. Moser, M. Gabay, P.A. Senzier, D. Jérome, K. Bechgaard, J.M. Fabre, Transverse transport in (TM)2X organic conductors: possible evidence for a Luttinger liquid. Eur. Phys. J. B Condens. Matter Phys. **1**(1), 39–46 (1998)
26. Z. Wang, W. Shi, R. Lortz, P. Sheng, Superconductivity in 4-Angstrom carbon nanotubes - a short review. Nanoscale **4**(1), 21–41 (2012)
27. B.D. Josephson, Possible new effects in superconductive tunnelling. Phys. Lett. **1**(July), 7–9 (1962)
28. K.K. Likharev, Superconducting weak links. Rev. Mod. Phys. **51**(1), 101–159 (1979)
29. V. Ambegaokar, A. Baratoff, Tunneling between superconductors. Phys. Rev. Lett. **10**(11), 486–489 (1963)
30. L.N. Bulaevskii, J.R. Clem, L.I. Glazman, A.P. Malozemoff, Model for the low-temperature transport of Bi-based high-temperature superconducting tapes. Phys. Rev. B **45**(5), 2545–2548 (1992)
31. C.K. Poole, H.A. Farach, R.J. Creswick, *Handbook of Superconductivity*, (Academic Press, Cambridge, 1999)
32. P.W. Anderson, Absence of diffusion in certain random lattices. Phys. Rev. **109**(5), 1492–1505 (1958)
33. A.A. Abrikosov, L.P. Gor'Kov, Superconducting alloys at finite temperatures. Sov. Phys. JETP **9**(1), 220–221 (1959)
34. P.W. Anderson, Theory of dirty superconductors. J. Phys. Chem. Solids **11**, 26–30 (1959)
35. S. Maekawa, H. Fukuyama, Localization effects in two-dimensional superconductors. J. Phys. Soc. Jpn. **51**(5), 1380–1385 (1982)
36. S. Maekawa, H. Ebisawa, H. Fukuyama, Upper critical field in two-dimensional superconductors. J. Phys. Soc. Jpn. **52**(4), 1352–1360 (1983)
37. A.M. Finkel'stein, Superconducting transition temperature in amorphous films. Pis'ma Zh. Eksp. Teo. Fiz. **45**(1), 37 (1987)
38. J.M. Graybeal, M.R. Beasley, Localization and interaction effects in ultrathin amorphous superconducting films. Phys. Rev. B **29**(7), 4167–4169 (1984)

39. A. Frydman, The superconductor insulator transition in systems of ultrasmall grains. Phys. C Supercond. Appl. **391**(2), 189–195 (2003)
40. D.B. Haviland, Y. Liu, A.M. Goldman, Onset of superconductivity in the two-dimensional limit. Phys. Rev. Lett. **62**(18), 2180–2183 (1989)
41. E. Bielejec, J. Ruan, W. Wu, Hard correlation gap observed in quench-condensed ultrathin beryllium. Phys. Rev. Lett. **87**(3), 036801 (2001)
42. T. Baturina, A. Mironov, V. Vinokur, M. Baklanov, C. Strunk, Localized superconductivity in the quantum-critical region of the disorder-driven superconductor-insulator transition in TiN thin films. Phys. Rev. Lett. **99**(25), 1–4 (2007)
43. V.F. Gantmakher, V.T. Dolgopolov, Superconductor - insulator quantum phase transition. Phys. Uspekhi **53**(1), 1–49 (2010)
44. M. Feigel'man, L. Ioffe, V. Kravtsov, E. Yuzbashyan, Eigenfunction fractality and pseudogap state near the superconductor-insulator transition. Phys. Rev. Lett. **98**(2), 027001 (2007)
45. M.V. Feigel'man, L.B. Ioffe, V.E. Kravtsov, E. Cuevas, Fractal superconductivity near localization threshold. Ann. Phys. **325**(7), 1390–1478 (2010)
46. I.S. Burmistrov, I.V. Gornyi, A.D. Mirlin, Enhancement of the critical temperature of superconductors by Anderson localization. Phys. Rev. Lett. **108**(1), 017002 (2012)
47. V.E. Kravtsov, Wonderful life at weak Coulomb interaction: increasing of superconducting/superfluid transition temperature by disorder. J. Phys. Conf. Ser. **376**, 012003 (2012)
48. G. Bianconi, Enhancement of Tc in the superconductor-insulator phase transition on scale-free networks. J. Stat. Mech. Theory Exp. **2012**(07), P07021 (2012)
49. M.V. Sadovskii, Superconductivity and localization. Phys. Rep. **282**(5), 225–348 (1997)
50. D.J. Thouless, Electrons in disordered systems and the theory of localization. Phys. Rep. **13**(3), 93–142 (1974)

Chapter 3
Experimental Methods

Measuring low-dimensional superconductivity requires advanced ultra-low temperature instruments and techniques. Operation of such apparatus is non-trivial: it is necessary to acquire a good understanding of their functional principles. This chapter aims to relay the essential information for the experimental work performed in this thesis.

First, our two custom-made cryogenic systems manufactured by Janis Research Company [1] are introduced: a dilution refrigerator and a variable temperature insert (VTI). The cryostats have been designed to operate with customized superconducting magnets, developed by Cryomagnetics [2] and customized by Janis Research Company. The two cryostats incorporate a tail corresponding to a long vertical insert which fits inside the superconducting magnets. This allows the sample area to sit at the center of the magnetic field. Operating superconducting magnets implies a large consumption of liquid helium (LHe): helium recovery and re-liquefiers are therefore matched up with the customized magnets.

The chapter then continues with the measurement apparatus. First are presented the electronics setup and functionalities necessary for the data collection during the measurements, with an explanation of the acquired parameters. Secondly, a general presentation of two fully automated systems by Quantum Design [3] is provided. These instruments were used for complementary transport and magnetization measurements: the physical property measurement system (PPMS) and the magnetic property measurement system (MPMS). The last part of the chapter is dedicated to the sample fabrication and contact preparation for the transport measurements.

3.1 The Dilution Refrigerator

The available temperature range of our cryogen-free dilution refrigerator lies between 4 K and ~20 mK. The pre-cooling is achieved by the mean of a He gas pulse tube refrigerator, composed of two main parts:

D. Ansermet, *Emergent Superconductivity in Low Dimensions*, Springer Theses,
https://doi.org/10.1007/978-981-13-2941-8_3

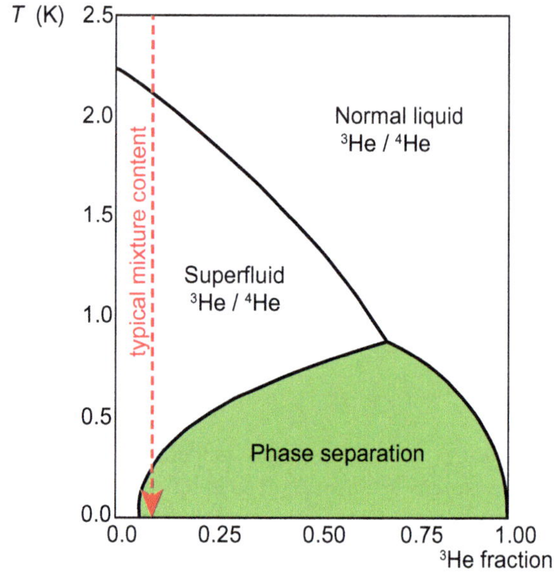

Fig. 3.1 Phase diagram of ^{3}He and ^{4}He gas mixture. The red arrow indicates the typical dilution refrigerator mixture content (~12% ^{3}He). Adapted from Pobell [4]

1. the pulse tube where the He gas is subject to smooth and periodic pressure variations and which possesses two heat exchangers at the entrance and exit for heat dissipation. At the high temperature side, a small orifice separates the heating and the cooling effect.
2. the regenerator situated at the cold temperature side and which consists of a porous material with large specific heat. The heat of compression at the end of the regenerator is dissipated to the surrounding via a heat exchanger.

To reach milliKelvin (mK) temperature, the system takes advantage of the enthalpy of mixing of two helium isotopes ^{3}He and ^{4}He.

^{4}He becomes superfluid at ~2.17 K [4], while ^{3}He remains a normal liquid at the temperature range considered here.[1] The temperature of the normal fluid-to-superfluid transition is lowered in ^{4}He diluted with ^{3}He (Fig. 3.1). At $T \sim 0.9$ K and ~67 % ^{3}He concentration, phase separation occurs: a pure ^{3}He (pure phase) coexists with a solution of superfluid ^{4}He and 6.6% ^{3}He (dilute phase).

The entropy, specific heat, and viscosity of ^{4}He are negligible below 1 K due to its superfluidity [5]. ^{4}He therefore constitutes an inert background only acting as a diluting medium for ^{3}He. From specific heat measurements [4], it has been shown that the enthalpy of ^{3}He in the dilute phase is larger than in the pure phase. Therefore, if ^{3}He is forced to migrate from the pure to the dilute phase, the temperature falls according to the enthalpy difference and a cooling power is obtained.

[1] ^{3}He becomes superfluid only at ~0.92 mK [4].

^{3}He has a finite solubility in ^{4}He even for $T = 0$ K. This is critical for the functioning principle of the dilution refrigerator: it ensures that the enthalpy exchange can take place even at the lowest operating temperatures.

3.1.1 Mixture Circulation and Condensation

The internal section of our dilution refrigerator is presented via a schematic in Fig. 3.2 and pictures in Fig. 3.3. In both figures the numbers correspond to the steps explained in the text. The names italicized in the text are indicated in the schematic.

1. The gas mixture comes from room temperature dumps through an injecting line (*return line*).
2. The gas mixture is precooled by the 60 K cold plate (*1st stage*) thermally connected by a highly flexible link to the pulse tube cold head's 1st stage, Fig. 3.3b.

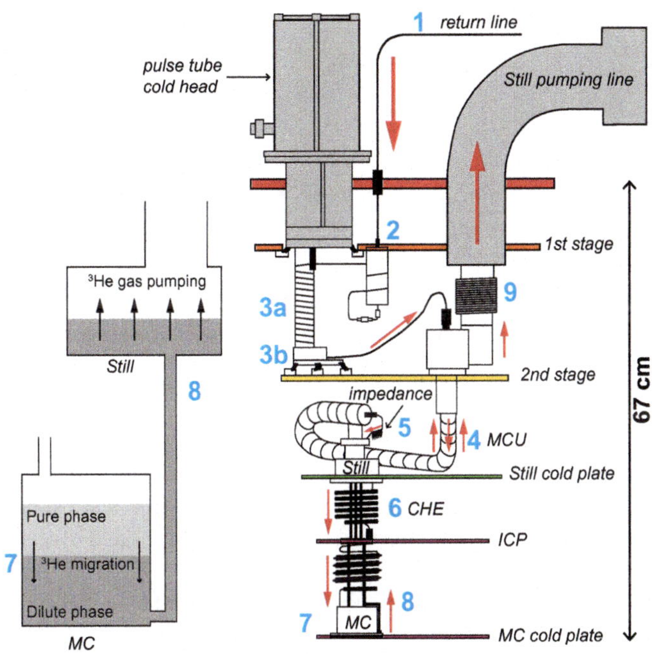

Fig. 3.2 Schematic of the mixture condensation and base temperature operating system. The blue numbers correspond to the enumerated steps in the text. The red arrows indicate the direction of gas mixture flow in the system: the gas is injected through the return line and pumped out via the Still pumping line. The zoom on the mixing chamber and still chamber depicts the ^{3}He migration mechanism within the condensate mixture

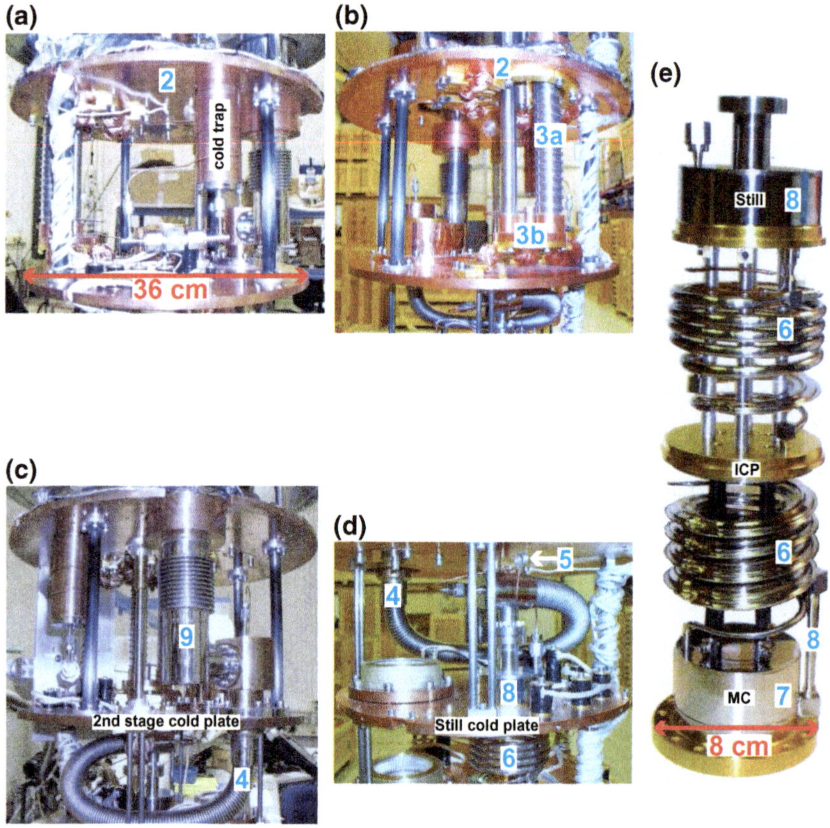

Fig. 3.3 Pictures of the mixture condensing system of our dilution refrigerator. The blue numbers correspond to the enumerated steps in the text. (**a, b, c**) The 1st and 2nd stage cold plates, with: **a** the cold trap, **b** the regenerator heat exchanger, and **c** the MCU and Still pumping line. **d** The 2nd stage and Still cold plates where the MCU, main impedance, CHE and Still chamber are visible. **e** The stage core including the CHE, ICP, MC, and Still chamber

3. (a) The gas mixture is further cooled using a regenerator heat exchanger, composed of a capillary soft-soldered to the 1st and 2nd stage of the pulse tube, Fig. 3.3b.
   (b) A 3 K cold plate (*2nd stage*) is thermally connected by a highly flexible link to the 2nd stage, Fig. 3.3b.

   This design allows the mixture to reach ~4 K at the 2nd stage. This is the signature of the Janis cryogen-free (dry) dilution refrigerator: pre-cooling to 4 K does not require a corresponding compressor and LHe bath, providing a more cost-effective system.
4. The gas mixture flows into the mixture condensing unit (*MCU*), a double-layered flexible hose where the cold ^{3}He gas exiting the system circulates around the outside, Fig. 3.3c, d.

5. The gas mixture is condensed by passing through a Joule-Thompson expander followed by a thin capillary with large flow resistance (*impedance*) with inner diameter ~0.25 mm and length ~30 cm, Fig. 3.3d. A 5 cm long secondary capillary follows the impedance to ensure full liquefaction.
6. The liquid mixture flows through a stainless steel spiral-in-tube continuous heat exchanger (*CHE*), Fig. 3.3d.
7. The liquid mixture accumulates in the mixing chamber (*MC*), where the phase separation occurs. An intermediate cold plate (*ICP*) minimizes heat leaks to the mixing chamber. The mixing chamber rests on its own cold plate, Fig. 3.3e. Because of its higher density, the dilute phase (^{4}He-rich) lies below the pure phase.
8. The liquid dilute phase flows to the Still chamber (*Still*) through a pipe at the bottom of the mixing chamber, Fig. 3.3e. The Still chamber contains a sharp-edged diaphragm to suppress superfluid film flow.[2]
   The Still chamber is used as a ^{3}He pump: ^{3}He has a much higher partial pressure than ^{4}He at the Still temperature, which leads to pure ^{3}He gas evaporation.
9. The cold ^{3}He gas circulates in the outside layer of the MCU before exiting the refrigerator through the *Still pumping line* which is of a large diameter (10 cm) to provide strong pumping power, Fig. 3.3c.
   The ^{3}He gas is finally re-injected through the return line to complete the mixture condensation loop.

Before initial condensation, a mechanical heat switch between the different cold plates provides an accelerated cooling. This heat switch is made of oxygen-free high thermal conductivity (OFHC) copper with a reduced diameter at each plate. A corresponding multi-contact thermal link is mounted at the 1st stage, 2nd stage, and Still cold plates. A clearance hole with a finger stock is mounted at the ICP cold plate. Finally, the heat switch terminates in a tapered end which matches a corresponding tapered receptacle on the MC cold plate.

The heat switch is manually engaged using a vertical linear manipulator. When starting the cool down from room temperature, the heat switch has to be fully engaged to ensure a good thermal contact between each stage, especially between the 2nd stage and the MC. The initial condensation will only take place with a heat switch fully engaged. However, the heat switch has to be removed once $T < 3$ K to reach mK base temperatures.

[2]Depending on their surface tension, ordinary liquids can creep up solid walls. This is observed for example with alcohol or petroleum. LHe is a superfluid and also possesses this property. However, in this case the behavior is restricted by a high critical velocity (about 20 cm/s). This means that LHe can very easily flow up the walls of the Still chamber.

3.1.2 Thermal Isolation and Radiation Shielding

Thermal isolation constitutes the bulk of the cryostat and includes isolation from heat conduction and convection, infra red (IR) radiations, and mechanical vibrations. A schematic and picture are presented in Figs. 3.4 and 3.5, respectively.

For heat isolation, the dilution refrigerator dewar is maintained under high vacuum ($\sim 10^{-6}$ mbar at room temperature). The vacuum is sealed by the outer vacuum chamber (*OVC*), an aluminum cylinder closed by two stainless steel plates. The thermal insulation of the electrical wires will be discussed in Sect. 3.1.5.

Each cold plate (1st stage, 2nd stage, Still, ICP and MC) is composed of OFHC copper with the Still, ICP and MC plates gold-plated. These provide thermal anchors for the mixture circulating system and measurement probe, and control the thermal gradient from the top at room temperature to the bottom at mK. The largest thermal gradient is between the 1st stage (60 K) and the 2nd stage (3 K). To achieve this massive gradient, the 2nd stage is attached to the 1st stage with carbon fiber supports, providing a good thermal insulation and strong mechanical resistance. The connections between 2nd stage, Still, and ICP are composed of stainless steel, and the link between the ICP and the MC cold plates can be of either carbon fiber or stainless steel.

Three IR radiation OFHC copper shields surround the mixture condensing system: these are attached on the 1st stage, 2nd stage and Still cold plates. A high vacuum is maintained between each shield to prevent conductive heating.

The dilution refrigerator incorporates a tail providing the geometrical configuration to operate with the customized superconducting magnets. The sample area is at the bottom of the tail, 89 cm below the MC cold plate (Fig. 3.4). The sample area is protected by the three IR shields, each of which incorporates a tail. The thermal contacts between the sample and the cryostat will be detailed in Sect. 3.1.6.

For additional thermal isolation, the 1st stage cold plate and shield (including the tail) are surrounded by multi-layer superinsulation. This is composed of highly reflective thin sheets of mylar: each layer of such superinsulation halves the black body radiation.

The pulse tube compressor inescapably brings noise and vibration. The cold head is installed on top of the cryostat (Fig. 3.5) and is pulsed by a remote motor connected to the compressor (Fig. 3.6). The pulse tube motor is installed remotely from the cold head using a 90 cm long flexible line to minimize vibrations, and electrical isolators are mounted between the cold head and the motor to minimize incident radio frequency (RF) noise and noisy grounds from the motor.

The equipment generating large noise or vibration (pumps, compressors, etc.) is located in the pump room and connected to the cryostat through the room's wall (Fig. 3.6). The measurement laboratory slab is mechanically disconnected from the pump room, isolating the measurement setups from the vibrations noise (Fig. 3.7). Any electrical ground-line noise generated by the pumps and vacuum gauges is blocked using a ceramic isolator on the Still pumping line. In addition, the measure-

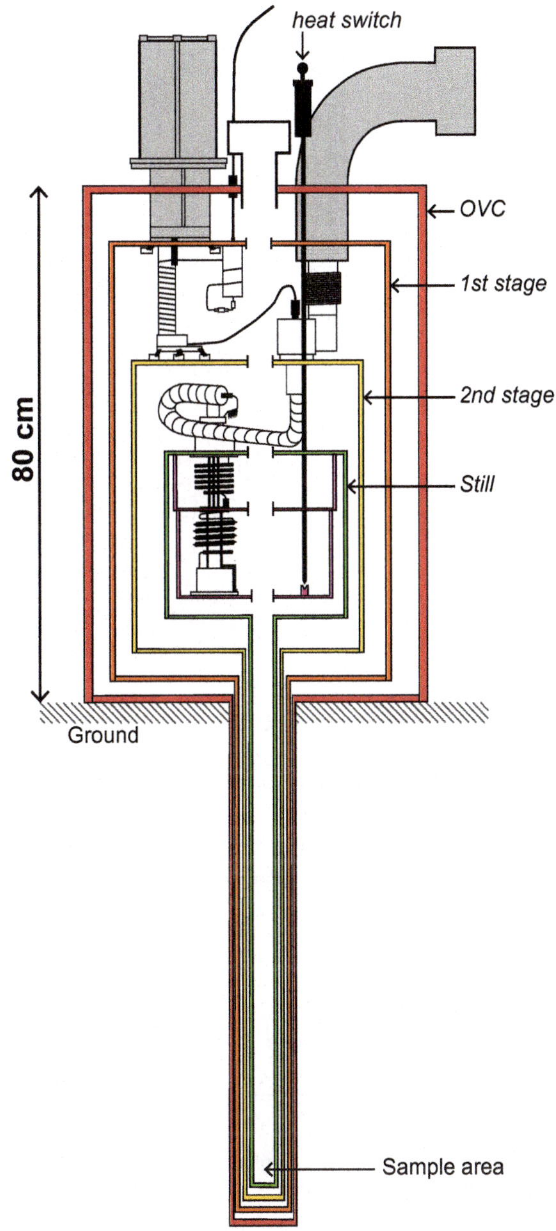

Fig. 3.4 Dilution refrigerator: thermal and RF shields of the cryostat chamber. Each shield is presented in a different color for ease of visualization. The cryostat incorporates a tail which sits either in a ground hole or inside the customized magnets

ment room is surrounded by a Faraday cage which is connected to a clean electric ground.

The gas mixture is stored in two dumps which can be isolated from the system with two valves when the cryostat is not operating. These dumps are fixed to the main control rack consisting of the various pressure gauges, valves, and control

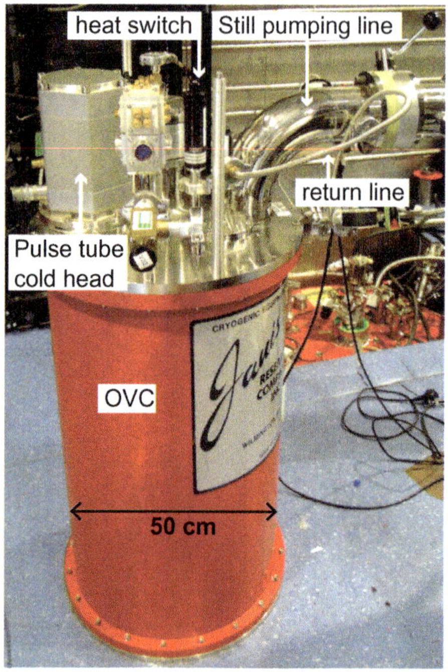

Fig. 3.5 Picture of the dilution refrigerator as viewed in the measurement room. The cryostat is sitting on top of a hole in the ground, to fit the length of the tail

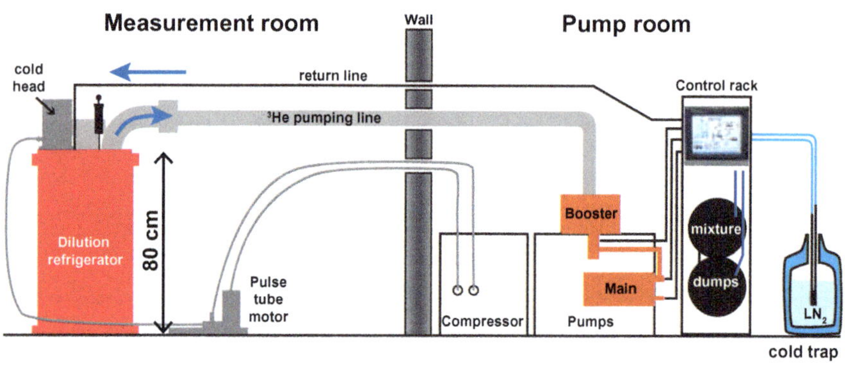

Fig. 3.6 Schematic of the complete dilution refrigerator system. The blue arrows indicate the direction of mixture circulation

computer, inside the pump room (Fig. 3.6). When operating the system with the control computer, the gas mixture is injected at low pressure from the dumps to the cryostat.

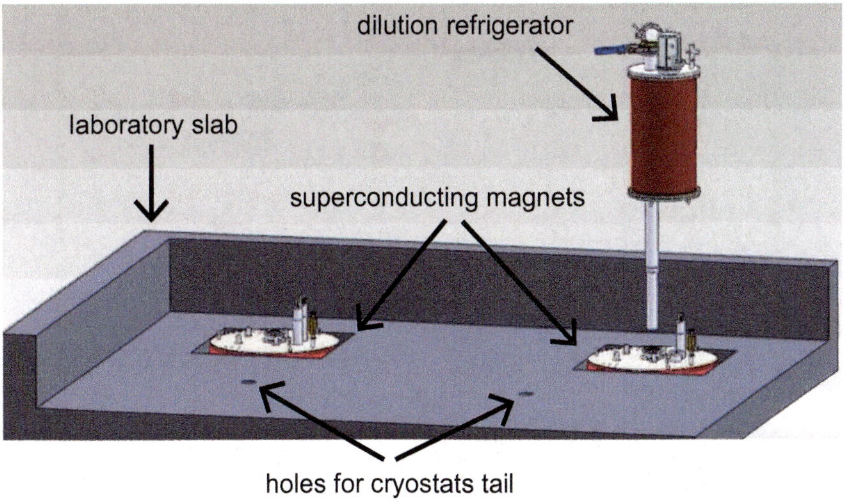

Fig. 3.7 Schematic of the laboratory slab. The cryostats (here the dilution refrigerator) can be moved from a superconducting magnet to another or can directly sit on the lab floor, where the tail fits into the hole in the laboratory floor. The laboratory slab is mechanically disconnected from the pump room. Snapshot of an animation from L. F. N. Ah Qune, PHYNE laboratory website [6]

3.1.3 Mixture Cleaning

The mixture is circulated in a closed loop. However, it is still necessary to consider an eventual impurity leakage which has the potential to block the main impedance during the cool down. Two systems are in place to clean the gas mixture.

First, before reaching the cryostat, the gas mixture passes through a LN_2 charcoal cold trap (Fig. 3.6) which cleans the gas from any impurity that can be absorbed by charcoal at $T \geq 77$ K. Before initiating a cool down, the gas mixture is cleaned through the cold trap for more than 24 h at a reduced flow. It is also essential to regenerate the cold trap regularly by isolating it from the mixture circulating loop, warming it up to room temperature and evacuating it to high vacuum.

Secondly, a refined gas mixture cleaning occurs at the 1st stage through a smaller charcoal cold trap mounted with a porous filter with 2 μm pore size preventing small charcoal particles from entering the refrigerator (Fig. 3.3a).

3.1.4 Mixture Circulation Control and Monitoring

The mixture gas circulating system is controlled using a Janis Research Company designed program installed in the control computer. A screen-shot of the operating

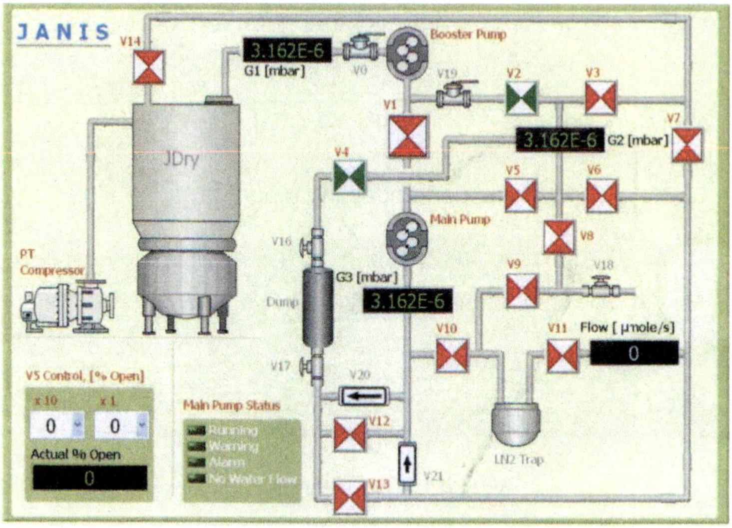

Fig. 3.8 Gas handling system as it appears on the computer program. The program allows remote control of the valves, pumps, and pulse tube compressor

program is shown in Fig. 3.8. The pulse tube compressor (PT compressor) and the two pumps can be turned on/off by the program.

A combination of electrically (G1, G2, and G3) and mechanically (G4, G5, G6 and G7) monitored gauges provide pressure readings for the full system. The mixture gas flow is also electrically monitored.

Several electrical valves (V1–V14) enable control of the gas mixture circulation using the program. During initial condensation, the gas flow is regulated through a variable aperture leak valve (V5). Additional mechanical valves (V15–V21) provide further safety measures such as preventing the gas mixture from circulating when the dilution refrigerator is disconnected.

3.1.5 Electrical Connections

It is essential to monitor and manipulate the temperature of each cold plate for successful mixture condensation and circulation. This is achieved via thermometers and heaters arranged as follows:

A calibrated Cernox^{TM} thermometer and a 25 Ω heater are installed on each of the 1st and 2nd stage cold plates; a RO600A ruthenium oxide temperature sensor with standard calibration and a 100 Ω heater are installed on the Still cold plate; a RO600A ruthenium oxide temperature sensor is installed on the ICP cold plate; a calibrated resistance JRS1000-05 ruthenium oxide temperature sensor and a 100 Ω heater are installed on the MC cold plate.

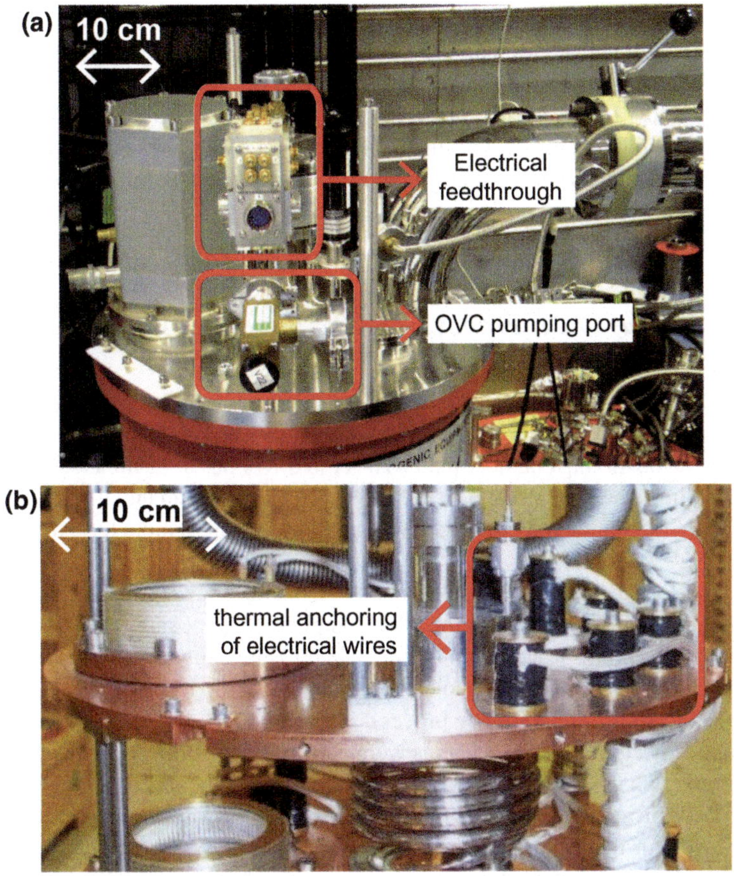

Fig. 3.9 **a** Electrical feedthroughs situated on the top plate of the refrigerator. Facing us are one 32-pin and 4 SMA feedthroughs. **b** Thermal anchoring of electrical wires at the Still cold plate. Such thermal anchors are also present on the 1st stage, 2nd stage, ICP and MC cold plate

The various heaters and thermometers are electrically connected via an external 32-pin feedthrough mounted on the OVC pumping port, situated on the OVC top plate (Fig. 3.9a). Additional electrical connections on the external port are provided to the user for measurement purposes: three 32-pin hermetic electrical feedthroughs and eighteen double-ended hermetic coaxial SMA feedthroughs.

Heat can flow through the electrical wires from the external feedthroughs. This may induce two undesired effects. First, warmer electrical wires would raise the cryostat temperature and affect the stability of the base temperature. Secondly, heat from the measurement cables might affect the sample under test (e.g. suppressing superconductivity).

To prevent heat flow, all electrical wires are anchored at each cold plate (Fig. 3.9b). Thermal anchoring involves winding the wires around a metallic bobbin and covering them with a thermally conducting resin.

Additionally, to prevent RF noise from entering the refrigerator, ferrite clip-on filters are installed on the cables outside the cryostat. Such ferrite beads are passive electrical components and dissipate induced high-frequency currents, thus suppressing high frequency noise.

3.1.6 Measurement Probes

Two measurement probes can be used with our customized dilution refrigerator: the cold finger and top loader probes. The cold finger consists of an internal silver insert fixed below the mixing chamber plate. The cold finger has to be mounted internally, which implies that the OVC and radiation shields have to be manually disassembled for sample installation and removal. Using this probe, the sample base temperature reaches $\geq$20 mK.

The top loader probe consists of an external vacuum space (load lock, Fig. 3.10) mounted on a flange situated on the OVC top plate. The vacuum is maintained by a double O-ring sliding seal and monitored by a pressure gauge. Various copper guide tubes located on the refrigerator cold plates prevent the top loader from being caught during the loading and unloading process.

In this project, most measurements were performed using the top loader, allowing us to prepare, change, and fix the sample and its electrical connections while keeping the dilution refrigerator under vacuum. However, the top loader probe needs careful management of the thermal contact and isolation to prevent heat diffusion. When this method is handled with precaution, the base sample temperature usually reaches $\sim$30 mK. In addition, the top loader provides a much wider measurement temperature range due to the weaker link to the mixing chamber.

Thermalization of the system is achieved through direct thermal contact between the probe and each cold plate. The probe's mating thermal stages are indicated in Fig. 3.10 by the names of their matching cold plates. Their sizes gradually decrease from the 1st stage to the mixing chamber. A good mechanical contact between the probe and the cold plates is achieved using the force generator gradually tightened during the cool down (Fig. 3.10). Good thermal contact and chamber vacuum are the keys to successful mixture condensation when using the top loader.

The sample stage is mounted at the bottom of the top loader (sample stage schematic and picture in Fig. 3.10). When the top loader is fully inserted, the sample sits at the bottom of the tail. The sample can be glued to the stage using GE varnish: a mechanically stable, thermally conducting, and electrically insulating epoxy. A temperature sensor and a 25 Ω heater are also mounted on the sample stage (sample stage schematic in Fig. 3.10).

The sample is connected using coaxial cables, composed of a conductive wire surrounded by a dielectric insulation, a conducting shielding and wrapped by a plastic

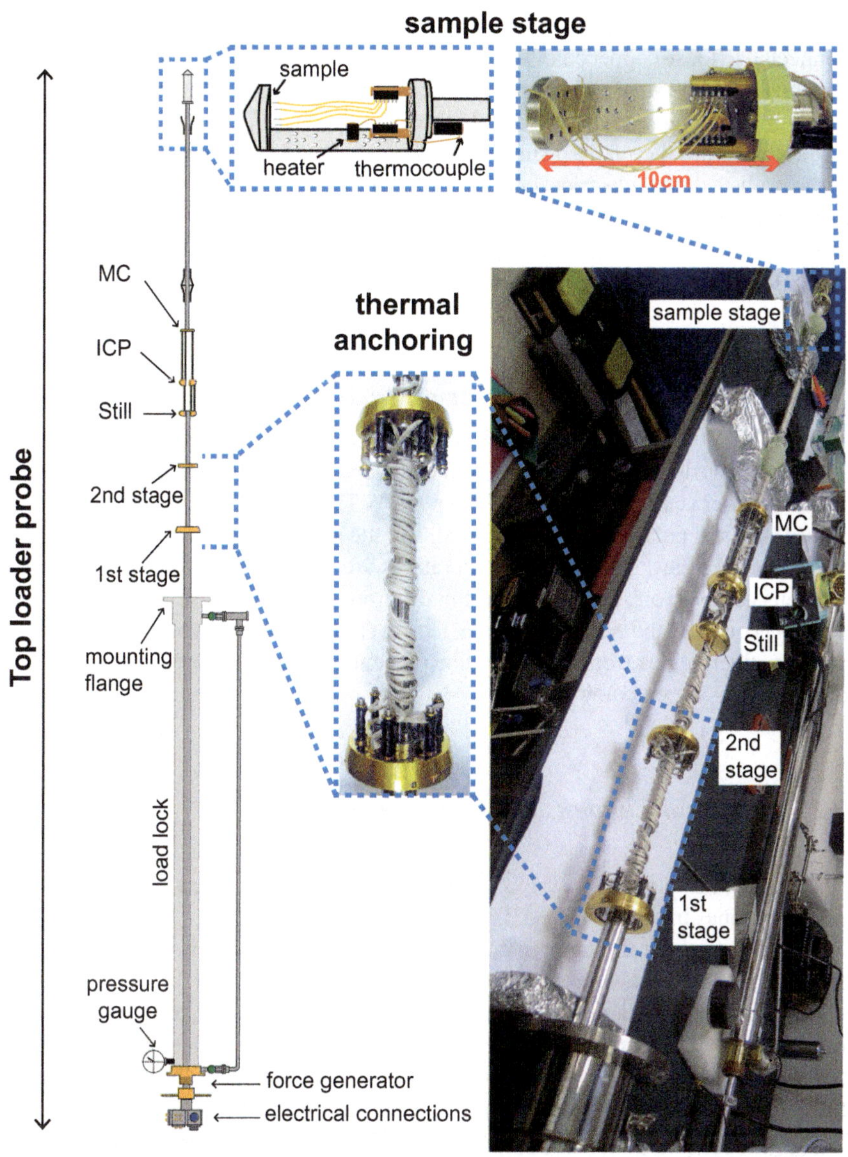

Fig. 3.10 The top loader probe used for the measurements in the dilution refrigerator. The mating thermal stages are indicated by the names of their matching cold plates. The **sample stage** and **thermal anchoring** are highlighted. The thermal anchoring is similar on all the mating thermal stages. The top loader has a total length of 363 cm, with the load lock mounting flange at the mid-point

jacket insulation. The conductive shielding is connected to the ground. The sample is connected to loose copper coaxial cables connected to small pins installed on the sample stage (sample stage schematic and picture in Fig. 3.10). The pins are connected to the probe stainless steel coaxial cables to reduce heat diffusion via the cables. The probe cables are thermally insulated and are connected to an hermetic feedthrough at the probe external electrical connections (Fig. 3.10). The unused feedthroughs are grounded using 50 Ω connectors to reduce RF radiation in the measurements. Heat diffusion through the wires is prevented via thermal anchoring at each mating thermal stage (thermal anchoring picture in Fig. 3.10).

3.2 The Variable Temperature Insert

The second customized cryostat used in this work was a VTI, whose main source of cooling power is a pulse tube refrigerator. Our VTI operates in the temperature range 1.6–300 K: the pulse tube refrigerator allows the system to reach ~3 K through the injection of He gas, while lower temperatures are reached by pumping a small LHe condensate reservoir. Comparing to the dilution refrigerator discussed earlier, this system is quicker and easier to operate. However, the finite LHe volume and the temperature fluctuations from the pulse tube prevent its use for extended measurements at fixed temperature below ~10 K.

3.2.1 Helium Condensation and Circulation

The various components described in the text are labeled by blue letters corresponding to the ones in Figs. 3.11 and 3.12.

The pulse tube refrigerator's 1st stage (A) base temperature is 57 K and the 2nd stage (B) ~3 K. A heat exchanger (C) is present between the 1st and 2nd stage, which consists of a stainless steel tube soldered onto the 2nd stage regenerator. The room temperature incoming gas is precooled at the 1st stage and is further cooled to approximately 4 K by flowing through the heat exchanger. Once the gas has reached the 2nd stage, it flows through a thin capillary (D) to then reach the vaporizer (E) which lies at the bottom of the tail. The gas condenses under high enough pressure (~7 PSI) and will start to accumulate at the vaporizer once its temperature is low enough (~4.7 K).

The LHe accumulates in the vapor cooled tube (F), a long tube of outer diameter 5 cm made of multiple sections of stainless steel and copper. The vapor cooled tube surrounds the sample tube (G), where the sample is inserted using a probe. A thermal link (H) made of copper straps connects the 2nd stage cold plate to the vapor cooled tube. Base temperature is reached by stopping the incoming He gas and gradually pumping on the LHe condensate inside the vapor cooled tube via a manual valve and a dry scroll pump.

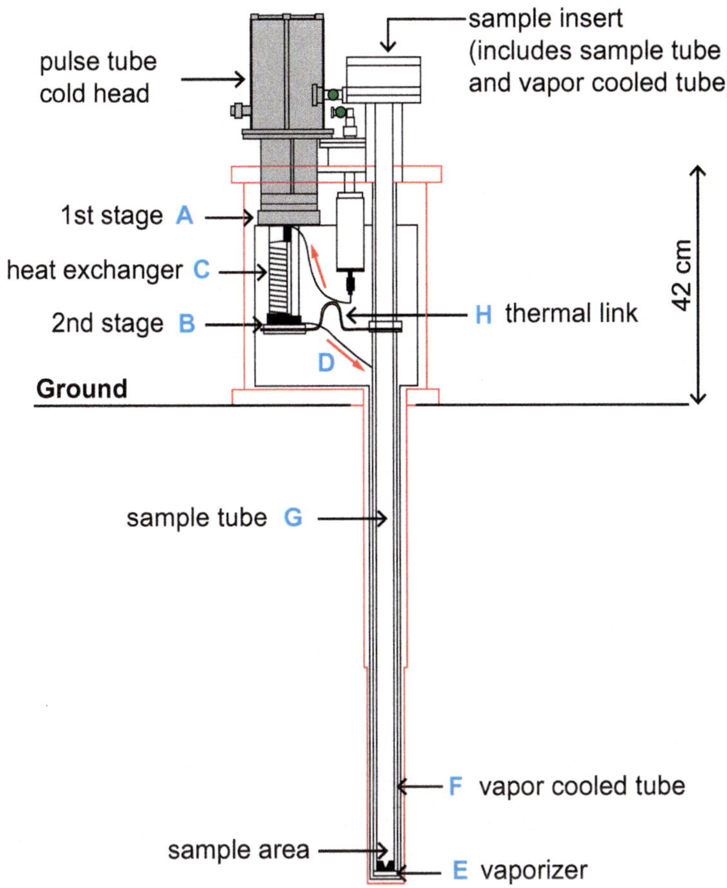

Fig. 3.11 Schematic of the VTI and its He liquefaction setup. The blue letters correspond to the components described in the text. The flow direction of the He gas is indicated by the red arrows

3.2.2 Thermal Isolation and Radiation Shielding

Thermal isolation is achieved by a high vacuum environment ($\sim 10^{-6}$ mbar at room temperature) in the OVC surrounding the cryostat. The OVC is made of an aluminum cylinder closed by stainless steel plates (Fig. 3.13). The bottom plate incorporates a stainless steel tail. The temperature gradient between the top plate and the vaporizer is achieved via a 50 K cold plate thermally connected by flexible copper braids to the pulse tube's 1st stage and a 3 K cold plate bolted onto the bottom of the 2nd stage (Fig. 3.12).

Inside the VTI, an IR radiation shield is attached to the 50 K cold plate. The shield incorporates a tail to protect the vapor cooled tube and the sample tube. The cold plate and tail are wrapped with mylar superinsulation to minimize black body radiation.

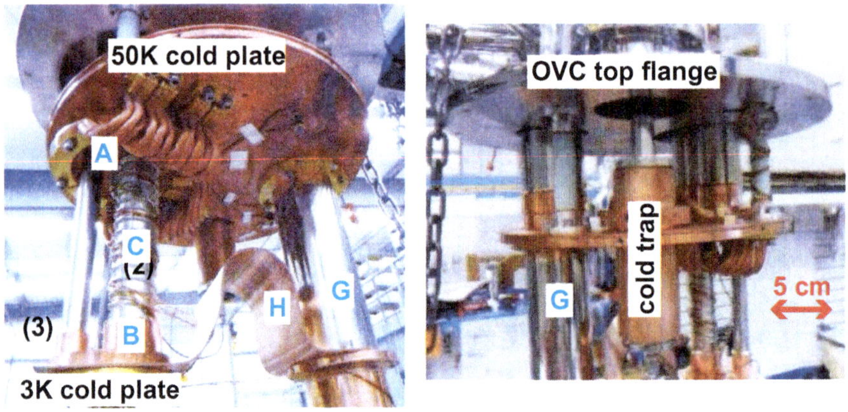

Fig. 3.12 Pictures of the He circulating system in the VTI. The blue letters correspond to the components in Fig. 3.11 and the description in the text

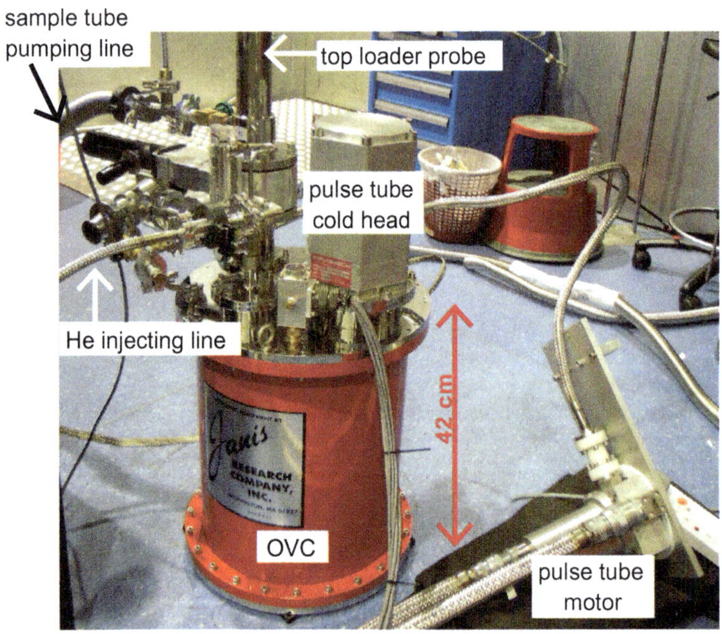

Fig. 3.13 Picture of the VTI as viewed in the measurement room. The cryostat is sitting on top of a hole in the ground to fit the length of the tail

To prevent mechanical noise from reaching the cryostat, the installation of the pulse tube refrigerator is identical to that of the dilution refrigerator: the pulse tube motor is installed remotely from the cryostat to minimize vibration (Fig. 3.13) and an electrical isolator is installed between the pulse tube cold head and the motor to reduce incident RF noise and ground fluctuations. Also, the compressor is stored in the pump room and is connected to the motor through the wall.

3.2.3 Helium Gas Cleaning

Before starting a cool down, room temperature He gas is flushed several times through the circulating system. Between each gas flush, the vapor cooled tube is pumped to vacuum. To prevent any remaining contaminant from blocking the circulation, a removable charcoal trap is installed on the 50 K cold plate (Fig. 3.12).

3.2.4 Electrical Connections

The electrical connections exit the cryostat through two 32-pin hermetically sealed feedthroughs shown in Fig. 3.14. One feedthrough is connected to the different heaters and thermometers inside the cryostat. A generic silicon diode thermometer and a 25 Ω heater are installed on both the 50 and 3 K cold plates. A calibrated Cernox^{TM} thermometer and a 25 Ω heater are installed at the bottom of the vaporizer.

Heat diffusion along the electrical wires is prevented by thermal anchors at the 50 and 3 K cold plates and wire wrapping around the heat exchanger, visible in Fig. 3.12.

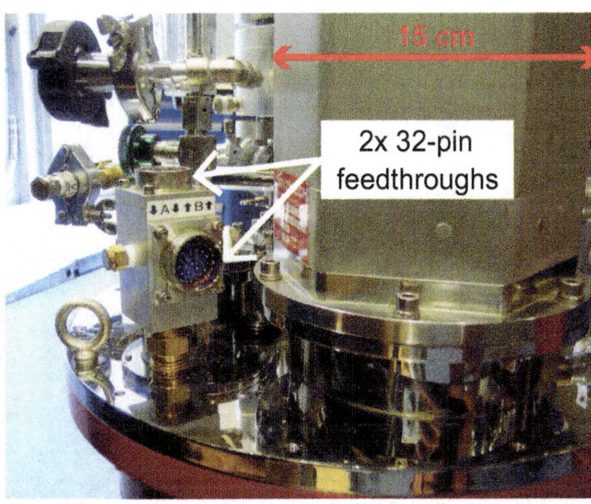

Fig. 3.14 Picture of the electrical feedthrough of the VTI: two 32-pin hermetically sealed feedthroughs provide connections to the heaters and thermometers of the cryostats, as well as extra wires for the user's measurements

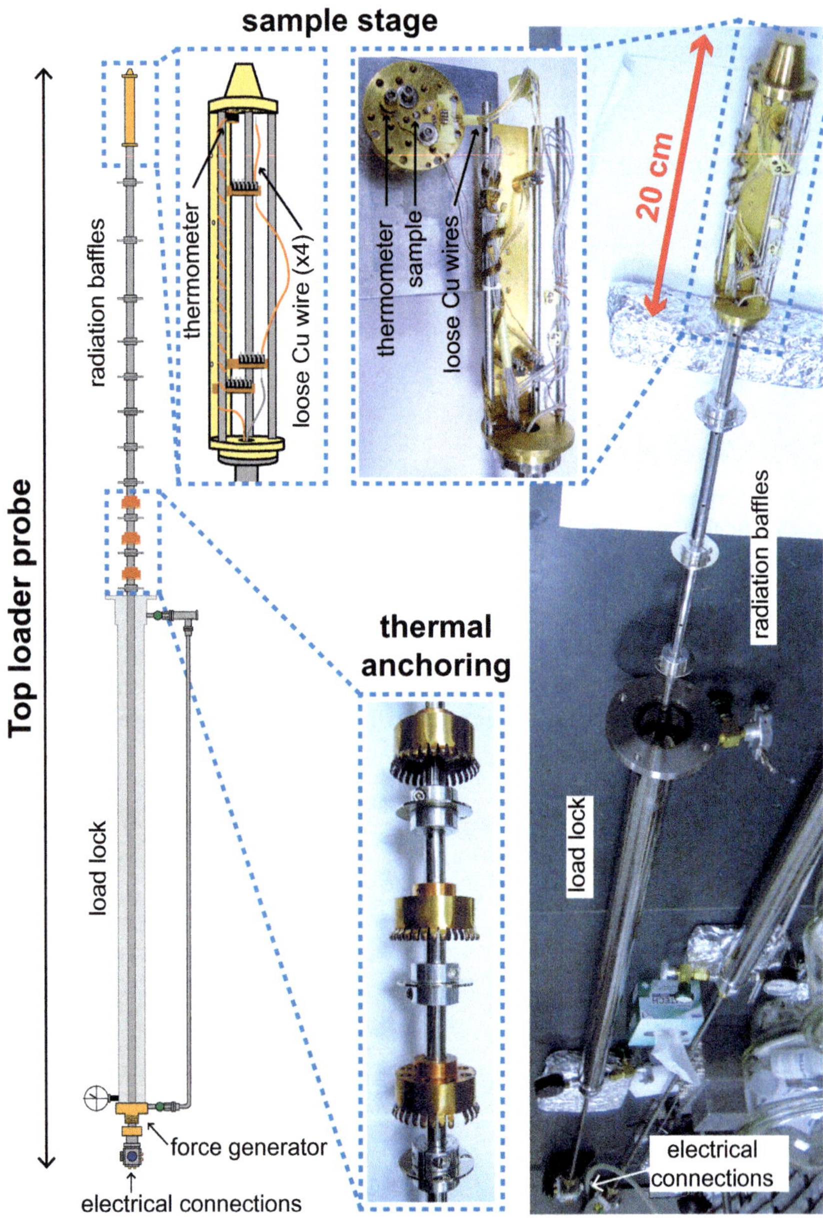

Fig. 3.15 The top loader probe used for the measurements in the VTI. The **sample stage** and the Be-Cu finger stock **thermal anchoring** are highlighted. On the schematic of the sample stage, only 1 loose Cu coaxial wire is depicted for clearer visualization, although 4 wires were actually connected

3.2.5 *Measurement Probe*

During measurements, the sample needs to lie at the bottom of the tail, just above the vaporizer (Fig. 3.11). The sample is inserted in the cryostat using a top loader probe (Fig. 3.15) which slides inside the sample tube. The sample tube is a vacuum tight tube welded to the vapor cooled tube at the top of the cryostat. The top loader is installed on the cryostat through its load lock mounting flange, connected to the sample tube's flange. The sample stage possesses a male tapered shape, which mates with a female tapered end at the bottom of the sample tube. A small force generator at the top of the probe ensures good mechanical contact between the probe and the sample tube, essential for a low base temperature at the sample. Thermal stabilization is maintained with Be-Cu finger stock thermal anchors at the 50 K plate and the 3 K plate (thermal achoring picture in Fig. 3.15). During normal operation, the sample temperature closely matches the temperature of the vaporizer.

A calibrated Cernox^{TM} thermometer is installed at the sample stage to separately monitor the temperature of the sample (sample stage schematic and picture in Fig. 3.15). The thermometer measures temperature in the 325–1.4 K range. The sample is fixed on the stage using GE varnish (sample stage picture in Fig. 3.15)

The probe incorporates two 32-pin hermetic feedthroughs and ten hermetically sealed double-ended SMA feedthroughs which exit at its top via the external electrical connections (Fig. 3.15). The 32-pin feedthroughs are connected to insulated manganin or insulated copper wires, while the SMA feedthroughs are connected to electrically floating copper or stainless steel coaxial cables.

During the measurements, the sample was connected to the SMA feedthroughs via the probe stainless steel coaxial cables (which possess a much lower thermal conductance than the copper wires) and to loose Cu coaxial wires (sample stage schematic and picture in Fig. 3.15). A system of two double-ended pins was used between the probe wires and the loose Cu wires to minimize thermal diffusion to the sample. This is an important step because in the VTI, no thermal anchoring of the wires is performed along the probe unlike the dilution refrigerator's top loader. All the unused feedthroughs were grounded using 50 Ω connectors to reduce noise in the measurements.

3.3 Superconducting Magnets

Magnets made of superconducting materials are an attractive solution to generate large magnetic fields without resistive heating. Also, as no dissipation occurs within the coil at constant field, a power supply rated at ∼1 kW (easily obtainable from a domestic 13 A socket) is sufficient to charge the magnet.

Table 3.1 Critical temperature T_c and field H_c of the two most common superconducting materials used in magnets

Material	T_c [K]	H_c [T] at 4.2 K
NbTi	9.8	12
Nb_3Sn	18.05	22.1

3.3.1 Superconducting Magnet Materials

Superconducting magnets operate below the T_c and H_c of the material from which they are made. Table 3.1 presents the two most common superconducting materials used in magnets: NbTi and Nb_3Sn.

From the values of T_c and H_c, it is necessary to operate the superconducting magnet inside a LHe bath. If the magnet is operated at temperatures or fields higher than the critical values, it will enter the normal resistive state. This will lead to a rapid heating of the system due to the large current flow through resistive wires, provoking a large boil off in the LHe bath. It is important to prevent such events and to protect the magnet and surrounding apparatus from the associated large energy dissipation and mechanical strain to the magnet system.

3.3.2 Stored Energy and Forces in a Magnet

The energy stored in a magnet depends on its geometry and operating conditions. The magnet being a solenoid, the amount of energy is:

$$W = \frac{1}{2}LI^2 \tag{3.1}$$

where L is the magnet inductance [H], I the operating current [A] and W the stored energy [J]. The typical magnets used in experimental laboratories can store a maximum energy of a few kJ to a few 100 kJ. When a magnet suddenly looses its superconducting state (quench), the stored energy is converted to heat in a few seconds. The windings need therefore to be protected using a copper matrix (twisted multi-filamentary superconducting/Cu wires) and diodes.

During operation, a large applied current is present in the magnet's winding generating a high magnetic field. The induced axial forces compress the winding and the induced radial forces expand the diameter of the magnet. Even small movements can lead to local heating due to friction and generate a quench. The wires are therefore impregnated in a strong epoxy layer, which restrains any movement. During the first few operations, the magnet needs to be trained, which consists of sweeping the field progressively to step values and returning back to zero between each step. Normally, once the magnet has been trained properly, no movement of the wires should occur.

Table 3.2 The design specifications of the custom-made superconducting magnet

Superconducting magnet	Vector	Solenoid
Central vertical field (at 4.2 K)	9 T	16 T
Central horizontal field (at 4.2 K)	4 T	–
Current at max. vertical field	50.5 A	114.2 A
Current at max. horizontal field	81.02 A	–
Field to Current ratio (vertical)	0.178 T/A	0.140 T/A
Field to Current ratio (perpendicular)	0.049 T/A	–
Total inductance L (vertical)	46.5 H	115.4 H
Total inductance L (horizontal)	52.4 H	–
Superconductor (Twisted Multi-filamentary wires)	NbTi/Cu	NbTi/Cu &Nb_3Sn

3.3.3 Magnet Specifications

The two superconducting magnets used during this project were: a 16 T solenoid and a 9 T/4 T vector magnet, both customized to operate with the dilution refrigerator and the VTI. The 16 T magnet consists of a single solenoid generating a vertical field; the 9 T/4 T vector magnet consists of a 9 T solenoid generating a vertical field and a 2 × 2 T split pair generating a horizontal magnetic field. The solenoid and split pairs of the 9 T/4 T vector magnet can be operated simultaneously up to a 3.5 T rotating vector sum. The technical details are presented in Table 3.2. The magnets are presented by schematics and pictures in Figs. 3.16 and 3.17, respectively.

In both systems, the magnets are surrounded by a LHe chamber which is thermally shielded by a LN_2 chamber and an OVC. The LN_2 reservoir has a capacity of 110 L and the LHe reservoir a capacity of 80 L. The levels of the LHe and LN_2 baths are monitored using level probes. Monitoring the level of the LHe bath is of primary importance as the magnet needs to be immersed in the liquid at all times during operation.

The nitrogen shielded LHe dewar opens to a neck on its top (Fig. 3.16). A helium gas venting adapter and a sliding seal are installed to the neck, to be sealed with the tail of the VTI or dilution refrigerator. The magnets are pre-installed on a plate at the bottom of the LHe bath, with the center of the magnet coil corresponding to the center of the sliding seal insert.

The cryostats are aligned by four male inserts (situated at their base) with four mating female support stand-offs on the magnet dewar top-flange (Fig. 3.17).

If no cryostat is installed in the magnet dewar, the dewar neck is closed with a blank flange. In both cases (with or without a cryostat), a slight over-pressure inside the dewar is maintained to prevent air from entering the system. The pressure is monitored with a gauge installed on the LHe fill port.

To monitor and optimize LHe loss during measurements, various flow meters are installed on the dewar's top flange: four flow meters are on the solenoid 16 T magnet and five flow meters are on the 9 T/4 T vector magnet (Fig. 3.17). The flow meters are connected to the helium fill port, the high current lead ports (one for the 16 T

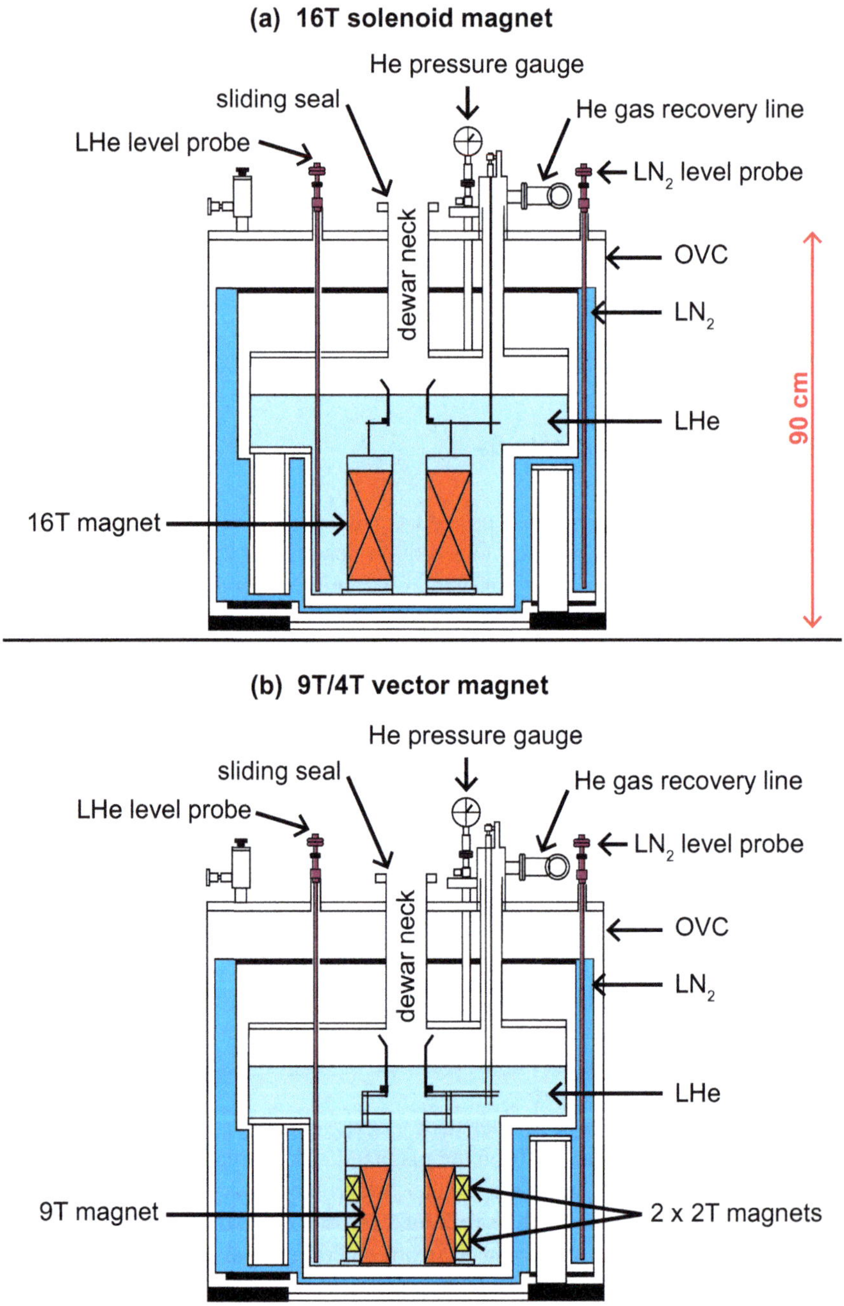

Fig. 3.16 Schematic of **a** the 16 T solenoid magnet and **b** the 9 T/4 T vector magnet systems. The magnets lie inside the LHe bath chamber, which is thermally isolated by a LN_2 bath and vacuum (OVC). The tail of the cryostats is inserted through the dewar neck

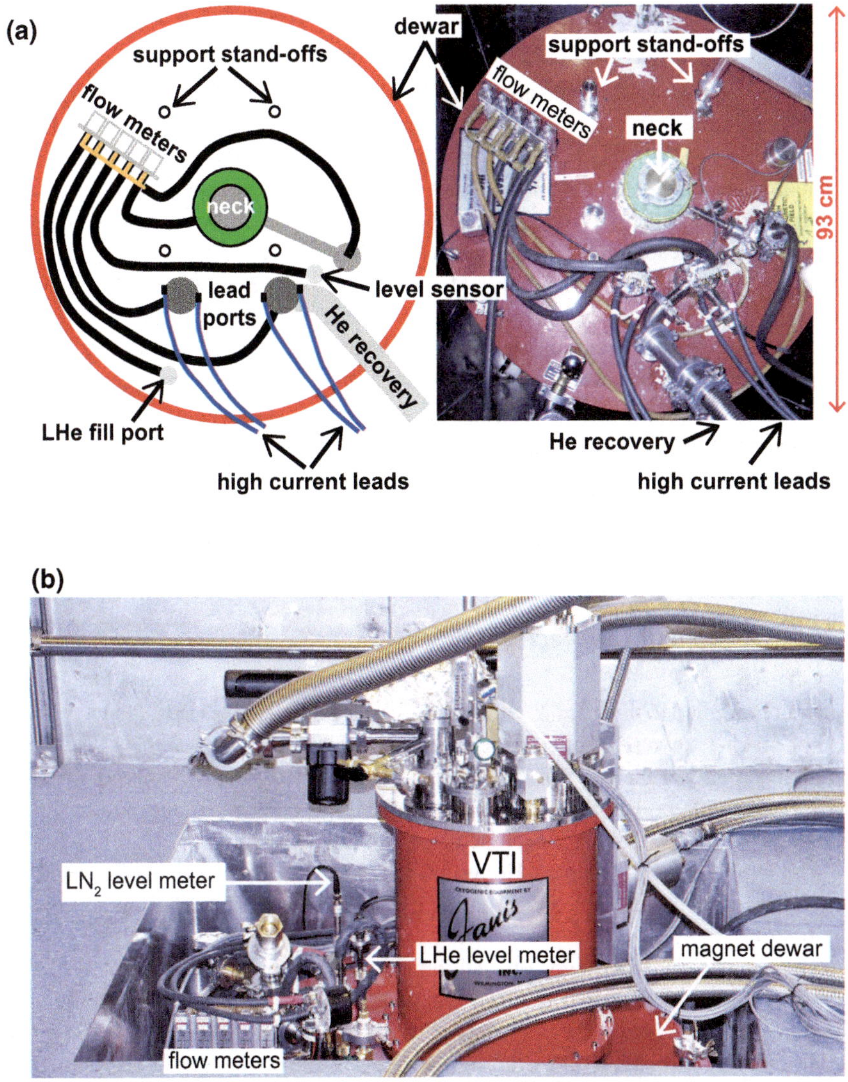

Fig. 3.17 The 9 T/4 T vector magnet as viewed in the measurement room. **a** Top view where a schematic highlights the key components. **b** side view where the tail of the VTI is inserted in the dewar neck

magnet, two for the 9 T/4 T vector magnet), the LHe level sensor port, and the dewar neck. The heat load introduced to the LHe reservoir varies at each port and also depends on the applied current through the current leads. The flow meters allow us to independently adjust the gas flow at each port.

3.3.4 Operating Modes

Once the magnet is charged, no further power is necessary to maintain the field (persistent mode). A current is therefore only required to vary the generated field (continuous mode). The persistent mode is achieved using a persistent switch, a superconducting wire across the input terminals of the magnet. A small heater is located close to the persistent switch. Before a current is injected into the magnet via the current leads, the heater must warm the persistent switch above its T_c to make it resistive. Once the desired field is generated, the heater is turned off and the persistent switch returns to its superconducting state. In this mode, the current flows in a closed loop inside the magnet and the current leads are no longer needed. No power is further required to maintain the magnetic field, which stays stable as long as the magnet is in LHe.

The persistent mode has two main advantages over continuous mode. First, turning off the power supply to the current leads strongly reduces the heat load on the LHe bath. Due to the high cost of LHe, it is important to minimize the boil-off. Secondly, the stability of the magnetic field is higher in the persistent mode.

3.3.5 Electrical Connections

Current is applied to the magnet via a pair of vapor-cooled brass leads. On the 9 T/4 T vector magnets, two pairs of current leads are installed: one pair for the 9 T solenoid magnet and one for the 2×2 T split pair magnet (Fig. 3.18). These leads are installed on separate vent ports and connected to a model 4G dual integrated power supply provided for the system and developed by Cryomagnetics.

The temperature inside the LHe reservoir is monitored using two Allen-Bradley carbon resistance thermometers installed on the magnet; one at the top and one at the bottom.

The electrical feedthroughs on each magnet dewar are as follows: for the user's personal requirements, one 32-pin hermetically sealed electrical feedthrough; for magnet wiring such as the voltage taps, the persistent switch heater(s) and the two Allen-Bradley carbon resistance thermometers, a 32-pin on the 9 T/4 T vector magnet (feedthrough *B* in Fig. 3.18) and a 19-pin hermetically sealed electrical feedthrough on the 16 T solenoid magnet. The LHe level sensor has a separate 4-pin hermetically sealed electrical feedthrough (feedthrough *C* in Fig. 3.18).

3.4 Helium Recovery and Re-Liquefying Systems

The magnets have a LHe storage capacity of 80 L and need to be refilled every 3–4 days, depending on the magnet operating mode (continuous or persistent). To avoid large expenses in LHe, a maximum of the evaporated gas is recovered and re-condensed. In our laboratory, two types of re-liquefier are used: a helium recovery and re-liquefying plant (Fig. 3.19a) and a small re-liquefier which can be mounted directly on the magnet (Fig. 3.19b).

The helium recovery and re-liquefying plant is installed outside the measurement room due to the noise generated and its large dimensions (Fig. 3.19a). The boil-off gas from the LHe magnet chamber is transferred to a large balloon which inflates with incoming gas. The gas is then compressed via a large compressor into empty bottles. The maximum recovery gas storage capacity is 555 L. The gas is purified by a LN_2 trap and then condensed using a pulse tube compressor. The liquid is finally stored in the re-liquefier dewar, which allows easy transfer to a portable dewar. At full working capacity, the plant can produce up to 21 L of LHe a day.

Additionally, a small re-liquefier can be mounted directly on top of the magnet (Fig. 3.19b). This provides a direct injection of LHe during measurements. A one-way gas pipe is connected to the recovery line (between the magnet and the balloon of the He recovery plant), which provides He gas to the small re-liquefier. A pulse tube compressor condenses the gas, which is directly re-injected into the LHe magnet chamber via a vertical pipe. The use of a one-way valve ensures a minimum pressure inside the re-liquefier, as a too low pressure might cause air to enter and freeze, blocking the pipes and the cold head. The daily production of the system depends on the boil-off rate of the magnet, and can be adjusted using a heater. In the case of the present work, the production was of the order of a few liters per day.

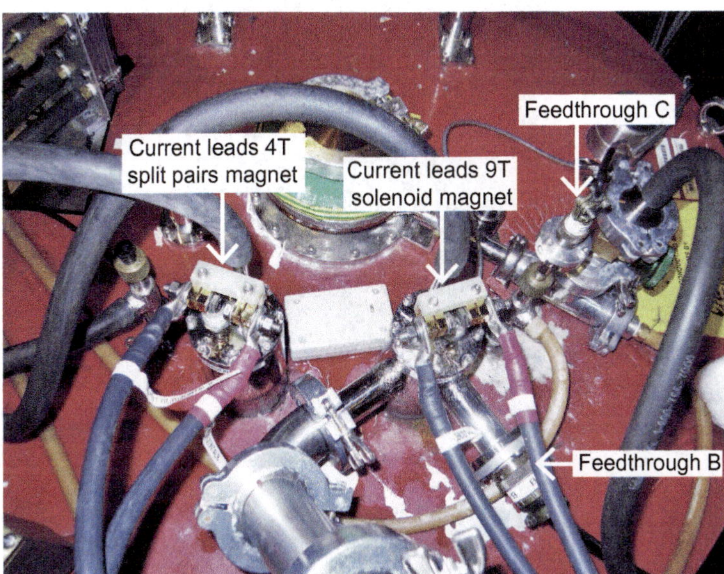

Fig. 3.18 Picture of the electrical connections on the 9 T/4 T vector magnet

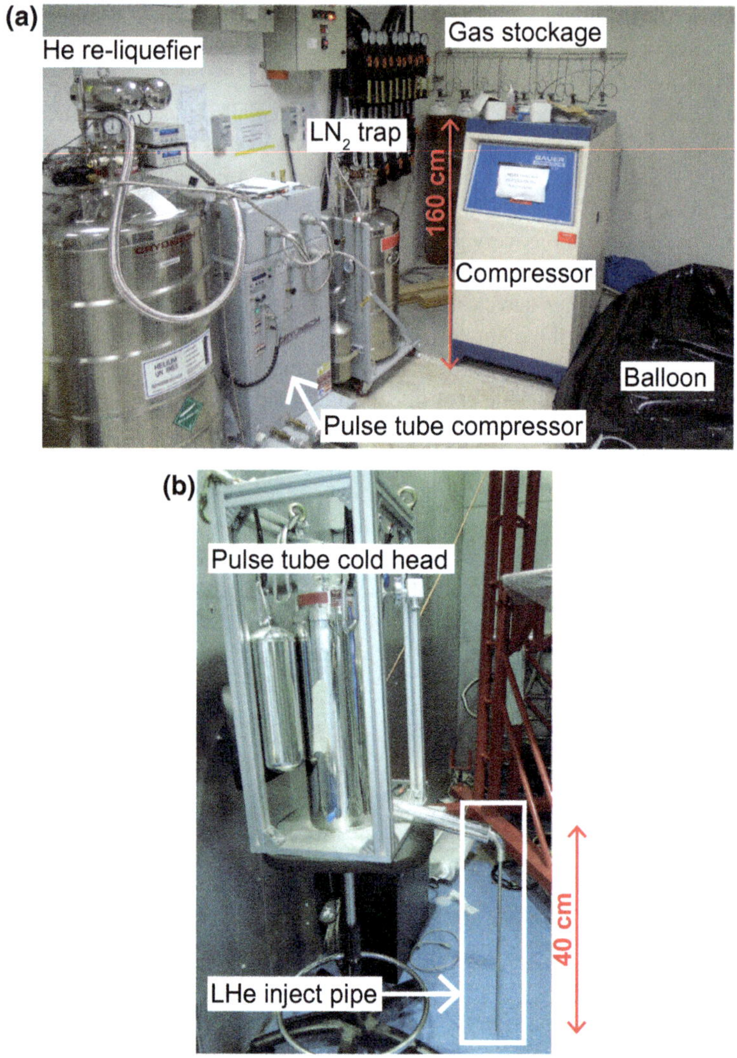

Fig. 3.19 The helium recovery and re-liquifying systems. **a** Main helium recovery and re-liquefying plant. **b** Small helium re-liquefier: the long LHe injecting pipe fits inside the LHe chamber of the magnet

3.5 Electronics

The samples were measured using the following electronic instruments (Figs. 3.20 and 3.21): a current source to apply an AC current (Keithley 6221 [7]), a lock-in amplifier to measure the frequency-dependent voltage signal (Stanford Research System 830 or 850 [8]), and two temperature controllers (Lakeshore models 336,

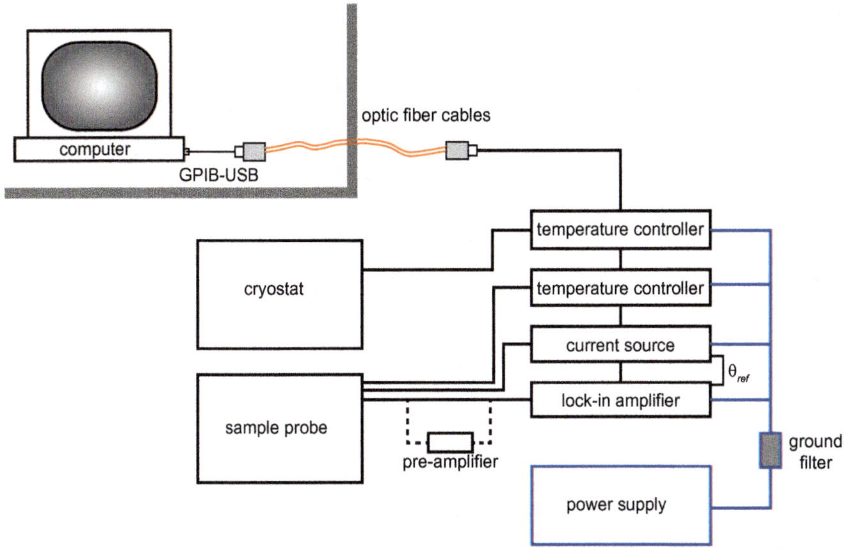

Fig. 3.20 Data acquisition setup for measurements in the dilution refrigerator and the VTI. In the case of very low resistance signals, a pre-amplifier was used to amplify the signal between the sample probe and the lock-in amplifier. A trigger link cable from the current source provides the reference phase θ_{ref} to the lock-in amplifier

340, or 370 [9]), one connected to the cryostat and the other to the measurement probe.

All the electronics were connected to a central computer via fiber optic cables. The connection to the computer was achieved via a General Purpose Interface Bus (GPIB) to USB connection (Fig. 3.20). The data were then acquired and recorded using Labview programs developed in our laboratory by A. Paré and S. He. The programs were personally adapted according to the measurement requirements and to improve data recording speed. The setup was similar for measurements in the dilution refrigerator (using two Lakeshore models 370) or the VTI (using one Lakeshore model 336 and one Lakeshore model 340).

3.5.1 Electronics Description

The current source: The current source is used to apply a current (DC or AC) to a sample. Ideally, it should output a constant current regardless of the sample or lead resistance. This system possesses an internal oscillator and provides a reference phase θ_{ref} to the lock-in amplifier.

The lock-in amplifier: The lock-in amplifier can be understood as a very sensitive, phase-dependent voltmeter. It detects AC signals using phase-sensitive detection which singles out very small measured signals at a given frequency from a large noisy background (up to thousands of times larger).

The sample is excited at an AC current with reference frequency ω_r. The lock-in detects the signal from the sample at ω_r as $V_{sig} \cdot sin\left(\omega_r t + \theta_{sig}\right)$, where V_{sig} is the signal amplitude and θ_{sig} is the signal phase. Then, the lock-in generates its own signal $V_L \cdot sin\left(\omega_L t + \theta_{ref}\right)$ using the phase reference θ_{ref} from the current source, where V_L and ω_L are the lock-in signal and frequency, respectively.

This develops a phase-sensitive detection (psd) as the lock-in will amplify the signal and multiply it by its reference:

$$V_{psd} = V_{sig} V_L sin\left(\omega_r t + \theta_{sig}\right) sin\left(\omega_L t + \theta_{ref}\right) \tag{3.2}$$

$$= \frac{1}{2} V_{sig} V_L cos([\omega_r - \omega_L]t + \theta_{sig} - \theta_{ref}) - \frac{1}{2} V_{sig} V_L cos([\omega_r + \omega_L]t + \theta_{sig} + \theta_{ref}) \tag{3.3}$$

A low pass filter singles out the signal for $\omega_r = \omega_L$, and the outputs from the phase-sensitive detector are two signals: one DC signal at the difference frequency $\omega_r - \omega_L = 0$ and one AC signal at the sum frequency $\omega_r + \omega_L = 2\omega_r$. The AC signal is extracted containing two main components:

1. the in-phase component X directly extracted from the previous equation with $\omega_r = \omega_L$:

$$V_{psd} = \frac{1}{2} V_{sig} V_L \cdot cos(\theta_{sig} - \theta_{ref}) \tag{3.4}$$

2. the out-of-phase component Y, or quadrature, using a second phase-sensitive detector to multiply the signal by the phase reference shifted by 90°:

$$V_{psd} = \frac{1}{2} V_{sig} V_L \cdot sin(\theta_{sig} - \theta_{ref}) \tag{3.5}$$

If $\theta_{sig} = \theta_{ref}$, X measures the signal while Y is zero.

The current reference frequency ω_r can be altered during the measurements: one needs to avoid the electricity grid frequency, in our case 50 Hz and its harmonics.

When an experiment is started, the phase reference is set as $\theta_{ref} = \theta_{sig}$. θ_{ref} is then kept constant and the evolution of θ_{sig} can be traced during the measurement.

The cryogenic temperature controller: To monitor and control the temperature down to LHe temperature and below, specific temperature sensors must be used. Various types of sensors are available: diodes, platinum resistance temperature detectors (RTDs), negative temperature coefficient RTDs or thermocouples. The appropriate sensor is selected depending on its temperature range. At the dilution refrigerator 1st and 2nd stages for example, the temperature sensors are Cernox^{TM}, thin-film negative temperature coefficient RTDs offering high sensitivity above 100 mK even in applied magnetic field. However at lower temperatures, Cernox^{TM} thermometers cannot be used due to self-heating and are replaced by ruthenium oxide sensors.

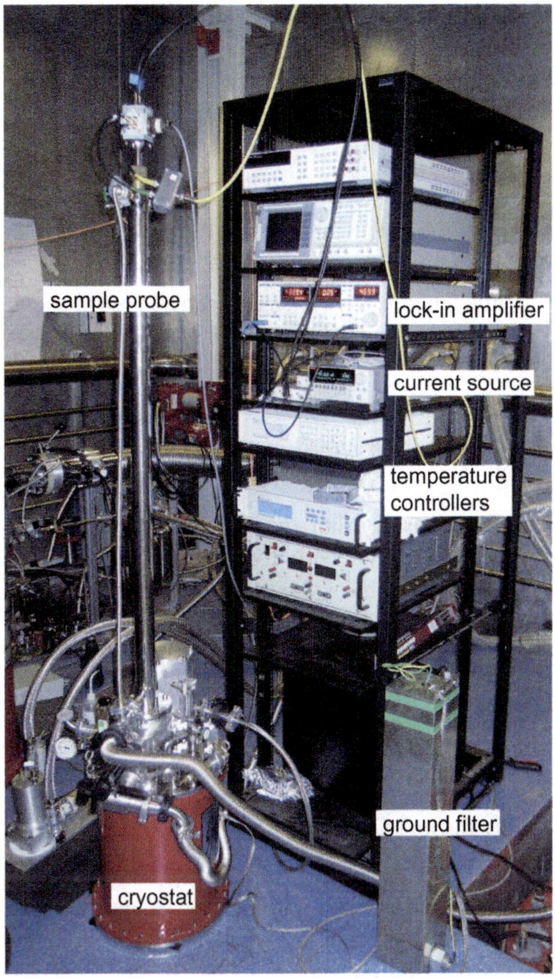

Fig. 3.21 Picture of the electronics setup used for the data acquisition in the transport measurements. Here, the electronics are connected to the sample probe of the VTI cryostat

Local temperatures can be manipulated using the temperature controller connected to small resistive heaters. These are composed of a resistive metal wire mounted on a small metallic bobbin and covered with a highly thermally conductive resin. They are essential to control the amount of condensed mixture inside the dilution fridge, LHe in the VTI and vary the sample temperature during sensitive measurements.

3.5.2 Electrical Noise Filtering

To prevent RF and high frequency noise from affecting the acquired data, a high performance ground filter manufactured by Holland Shielding Systems [10] was introduced between the experiment earth and our electronics to remove all ground current (Fig. 3.21). It consists of a low-pass LC circuit and filters signals above a certain cutoff frequency. It provides an insertion transmission loss of 100dB in the frequency range of 14 kHz–40 GHz.

3.6 Automated Measurement Systems

Two fully automated systems were also used to acquire the experimental data. These systems are developed by Quantum Design [3] to be auto-regulated and user friendly.

3.6.1 The Physical Property Measurement System

The PPMS used in this work possesses a 1.7–300 K temperature and −14–14 T magnetic field range. It is composed of three main parts (Fig. 3.22): 1. the measurement dewar which consists of a LHe dewar, a superconducting magnet and a vacuum sample chamber; 2. the control tower which contains all the electronic instruments necessary for data acquisition; 3. a helium re-liquefier (Cryomech PT410) which recovers the dewar helium boil-off and liquefies injected He gas. The PPMS provides quick and accessible transport, magneto-transport and susceptibility measurements.

In this work, the PPMS was mainly used for transport measurements using the AC transport (ACT) option.

AC Transport Option

The ACT option provides precise measurements using a current source and a precision voltmeter (lock-in amplifier) incorporated in the control tower. The resolution of the current source is 0.02 μA and the maximum applied current is 1 A. The voltmeter possesses a similar resolution and range. The available frequency range of the AC current source is from 1 Hz (DC limit) to 1 kHz.

The ACT option supports four types of electrical transport current measurements: resistivity, Hall coefficient, I-V curve and critical current. Basically, I-V curve and critical current are variants of the resistivity measurement and all require the same type of contact connections to the sample. For the Hall measurement, however, a different contact configuration has to be used.

In this work, the ACT option was used to measure resistivity and I-V curves at frequency 1–800 Hz. The data acquired presented a high accuracy with a noise level < 5 nV at $I = 10$ μA (in the dilution refrigerator or VTI, noise level < 20 nV at $I = 10$ μA).

The system provides a very user-friendly approach to measurements. The sample simply needs to be glued with GE varnish and connected to a small PPMS puck using 4-probe contact setup (Fig. 3.23). The sample puck is then inserted into the measurement dewar and the contacts are connected to the internal wires via bottom

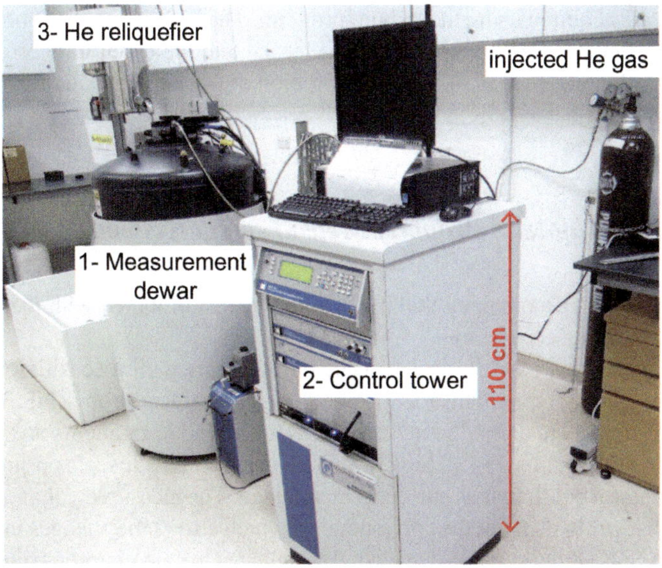

Fig. 3.22 Picture of the PPMS complete system

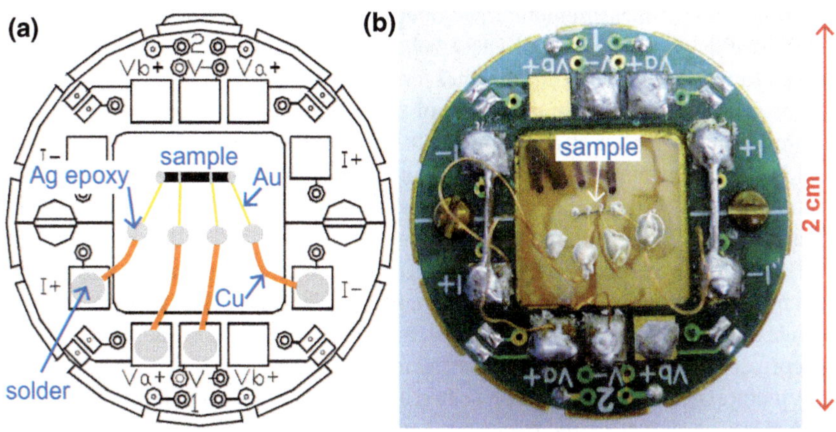

Fig. 3.23 The PPMS AC Transport sample puck: **a** schematic and **b** picture. The sample is glued in the center using GE varnish and the sample contacts are soldered to the puck's contact pads

pins. Data are acquired using the in-built program. The lock-in amplifier provides the in-phase component X, the full signal $\sqrt{X^2 + Y^2}$ and the quadrature error $Y \times \delta\phi$ where $\delta\phi$ is the tolerance in the phase calibration of the lock-in, usually in the order of 0.1 degrees [3].

3.6.2 The Magnetic Property Measurement System

The MPMS is a self-contained SQUID magnetometer combined with the necessary electronics for data acquisition.

This apparatus is used to detect magnetic moments with a sensitivity down to $\leq 10^{-8}$ emu. The sample is inserted inside the measurement probe (Fig. 3.24a) and oscillates inside a superconducting magnet. The magnetic moments are picked up by the SQUID capsule (Fig. 3.24b), which consists of two Josephson junctions in parallel, where a constant bias current is maintained. The measured voltage oscillates with the changes in phase at the two junctions, which allows the changes in magnetic flux in the SQUID loop to be detected. The induced voltage is proportional to the magnetic moment of the sample: by counting the oscillations, the magnetic moment can be extracted.

During sample measurement, raw voltage readings are taken by the SQUID as a function of the sample position. These voltage readings are performed several times at each sample position to improve measurement resolution.

The sample is inserted in the center of the SQUID using a drinking straw and is secured using a smaller straw segment. It is preferable not to use any glue, which could add an unwanted magnetic signature from trapped moisture or oxygen, or from impurities in the glue itself. The straw is then fixed to the sample rod, which is inserted inside the measurement probe (Fig. 3.24a).

The MPMS is calibrated by measuring a palladium reference sample of cylindrical dimensions 3 mm in diameter and 3 mm in height. The calibration is performed over a range of magnetic fields, where the system calibration factors are adjusted to obtain the correct magnetic moment of the standard sample.

A picture of the full MPMS is presented in Fig. 3.25. The system is divided into two main parts: on the left is the measurement dewar which includes the measurement probe and the LHe dewar; on the right is the console cabinet which includes the superconducting magnet power supply, the power distribution unit, the gas control unit, the vacuum pump, and all the necessary electronics to extract and process the data.

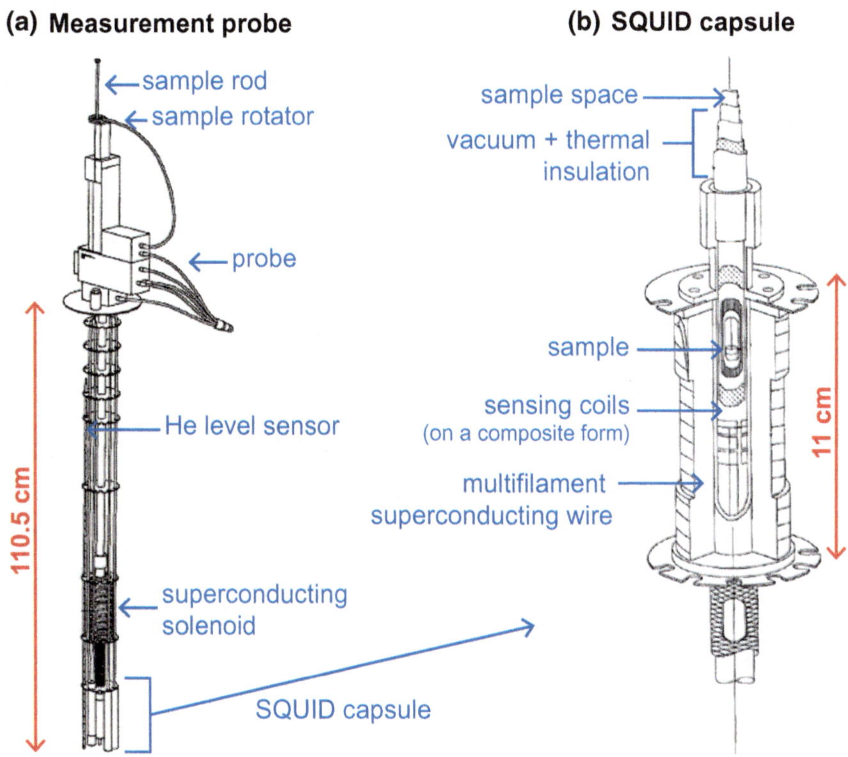

Fig. 3.24 **a** The MPMS measurement probe. **b** The SQUID capsule located at the bottom of the probe

3.7 Sample and Contacts Preparation

The samples measured in this work were single crystals of $Na_{2-\delta}Mo_6Se_6$ (where $\delta > 0$). In this section, only the preparation of the samples and contacts is described. Material structure and properties are covered in Chap. 4.

3.7.1 Sample Growth

The samples were grown at the Institutes des Sciences Chimiques de Rennes (CNRS, France) by our collaborators D. Salloum, P. Gougeon and M. Potel.

$Na_{2-\delta}Mo_6Se_6$ crystals were grown using a solid-state synthesis procedure [11–14]. The precursor materials were $MoSe_2$, InSe, Mo and NaCl, all in powder form. Before use, the Mo powder was reduced under H_2 gas flowing at 1000 °C for ten hours, in order to eliminate any trace of oxygen. The $MoSe_2$ was prepared by reacting Se with H_2-reduced Mo in a ratio 2:1 inside a purged, evacuated and

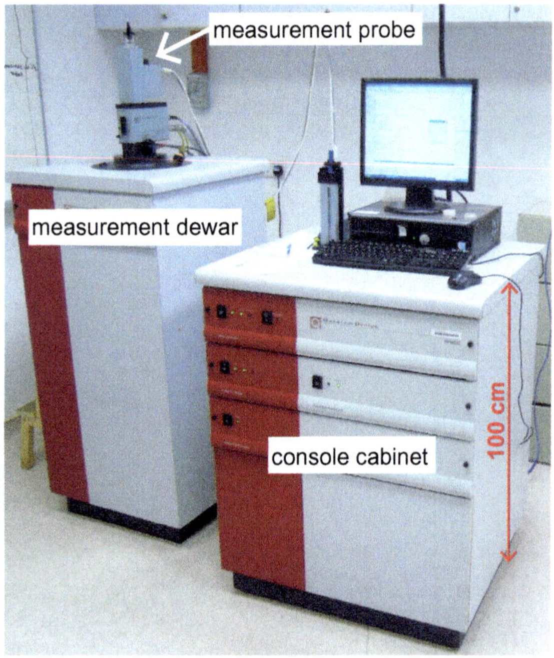

Fig. 3.25 Picture of the MPMS complete system

flame-baked silica tube (with a residual pressure of $\sim 10^{-4}$ mbar argon), which was then heated to $\sim 700\,^\circ$C for two days. InSe was synthesized from elemental In and Se in an evacuated sealed silica tube at 800 °C for 1 day.

Powder samples of $Na_{2-\delta}Mo_6Se_6$ were prepared in two steps. First, $In_{2-\delta}Mo_6Se_6$ was synthesized from a stoichiometric mixture of InSe, $MoSe_2$ and Mo, heated to 1000 °C in an evacuated sealed silica tube for 36 h. Second, an ion exchange reaction of $In_{2-\delta}Mo_6Se_6$ with NaCl was performed at 800 °C, using a 10% NaCl excess to ensure total exchange [12]. All starting reagents were found to be monophase on the basis of their powder X-ray diffraction patterns, acquired using a D8 Bruker Advance diffractometer equipped with a LynxEye detector ($CuK\alpha_1$ radiation) [12–14]. Furthermore, in order to avoid any contamination by oxygen and moisture, the starting reagents were kept and handled in a purified argon-filled glovebox.

To synthesize single crystals, a $Na_{2-\delta}Mo_6Se_6$ powder sample (of mass ~5 g) was cold-pressed and loaded into a molybdenum crucible, which had previously been out-gassed at 1500 °C for 15 min under a dynamic vacuum of $\sim 10^{-5}$ mbar. The Mo crucible was subsequently sealed under a low argon pressure using an arc-welding system. The $Na_{2-\delta}Mo_6Se_6$ powder charge was heated at a rate of 300–1750 °C, held at this temperature for 3 h, then cooled at 100 °C/h down to 1000 °C and finally cooled naturally to room temperature within the furnace.

The resulting samples are dark, slightly shiny, needle-like single crystals of approximate mass 10–150 μg, length 0.1–2 mm and diameter 5–200 μm. Scanning electron microscope (SEM) images of $Na_{2-\delta}Mo_6Se_6$ crystals are shown in Fig. 3.26.

The samples are extremely fragile need therefore to be handled with great care during cleaning and contact preparation (Fig. 3.26a). The crystals selected for the measurements were all intact (resembling the crystal shown in Fig. 3.26b).

Once grown, the crystals are very stable under atmospheric conditions and do not suffer from any oxidation or impurity diffusion. Their lifetime during measurements is determined by their high fragility and sensitivity to thermal cycling. Nevertheless, the crystals always present a certain amount of Na vacancies during their synthesis ($\delta > 0$), which constitutes an intrinsic disorder. Crystal characterization revealed Na deficiencies up to 13% (Sect. 4.3).

3.7.2 *Electrical Contact Preparation*

The samples selected for transport measurements had to be of a minimum length (a few hundred μm) for optimized contact size and separation. They were also chosen for their homogeneous surface and shape (constant cross-section). Before contact deposition, the crystals were cleaned with dilute hydrochloric acid (few seconds), then acetone (30 min) and ethanol (30 min). This cleaning procedure ensured an uncontaminated surface from their growth in a Mo crucible and storage. Next, contacts were deposited to probe a straight, undamaged section of the crystal. The

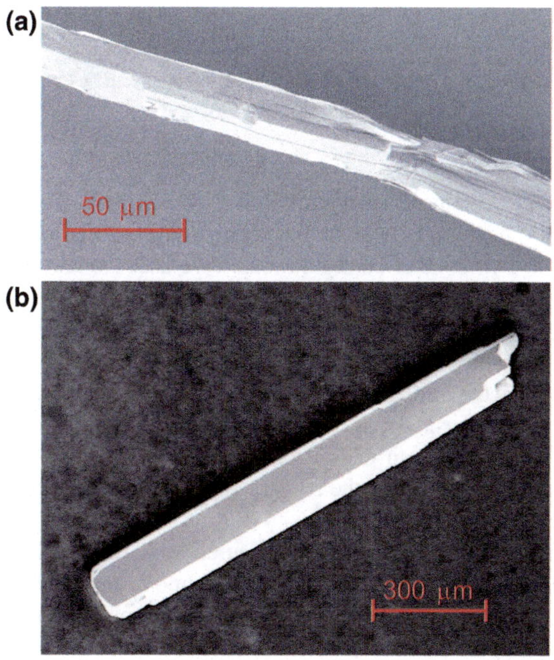

Fig. 3.26 SEM picture of $Na_{2-\delta}Mo_6Se_6$ single crystals. **a** damaged crystal due to manipulation. **b** intact crystal typically used for transport measurements [14]

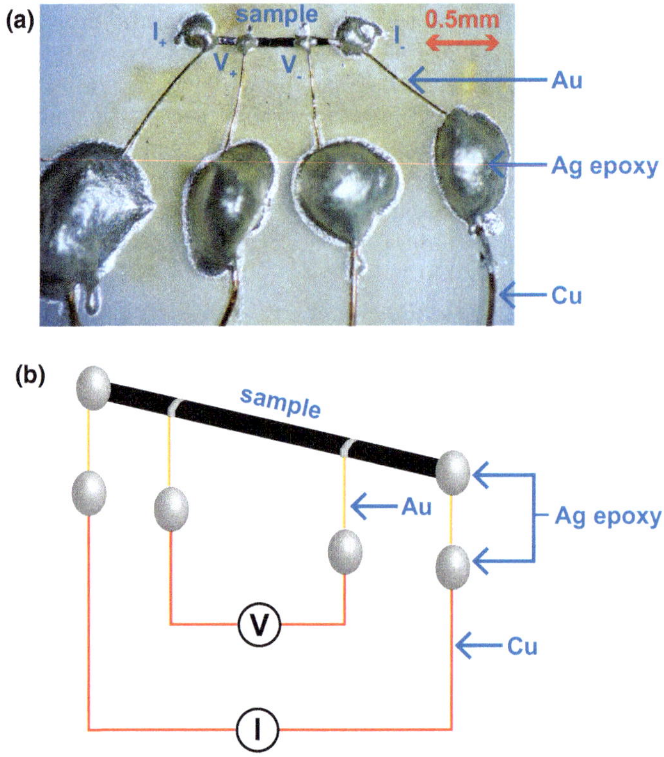

Fig. 3.27 4-Probe longitudinal contacts on $Na_{2-\delta}Mo_6Se_6$ samples. Silver (Ag) epoxy, gold (Au) and copper (Cu) wires are indicated. **a** Picture taken by optical microscope. **b** The four contacts are indicated as set for the 4-probe measurement technique

substrates used for mounting the samples were $10 \times 10\ mm^2$ sapphire plates, which provide good electrical insulation and good thermal conductivity.

Two contact configurations were used: longitudinal, to probe the current along the main 1D axis of the crystal (Fig. 3.27), and transversal, to probe the current in the direction perpendicular to the main axis (Fig. 3.28).

In the longitudinal configuration, four Au contacts were first sputtered onto the surface by DC magnetron sputtering for 100 sec with $I = 30$ mA using hand-made Al foil mask. Then, a 50 μm Au wire was glued to each Au pad using Ag epoxy, a 2-component air-drying paste (Epotek E4110). The 50 μm Au wires are very fragile and were fixed to the substrate with Ag epoxy and connected to 0.1 mm insulated copper wires (Fig. 3.27).

The contacts were finally cured at 70 °C for 3 h, as recommended by the epoxy manufacturer. Each end contact (I_+ and I_- respectively) thoroughly coats the full tips of the crystal to ensure that the current applied passes through the entire crystal. The voltage contacts (V_+ and V_-) cover the full visible width of the crystal for optimum measurement accuracy.

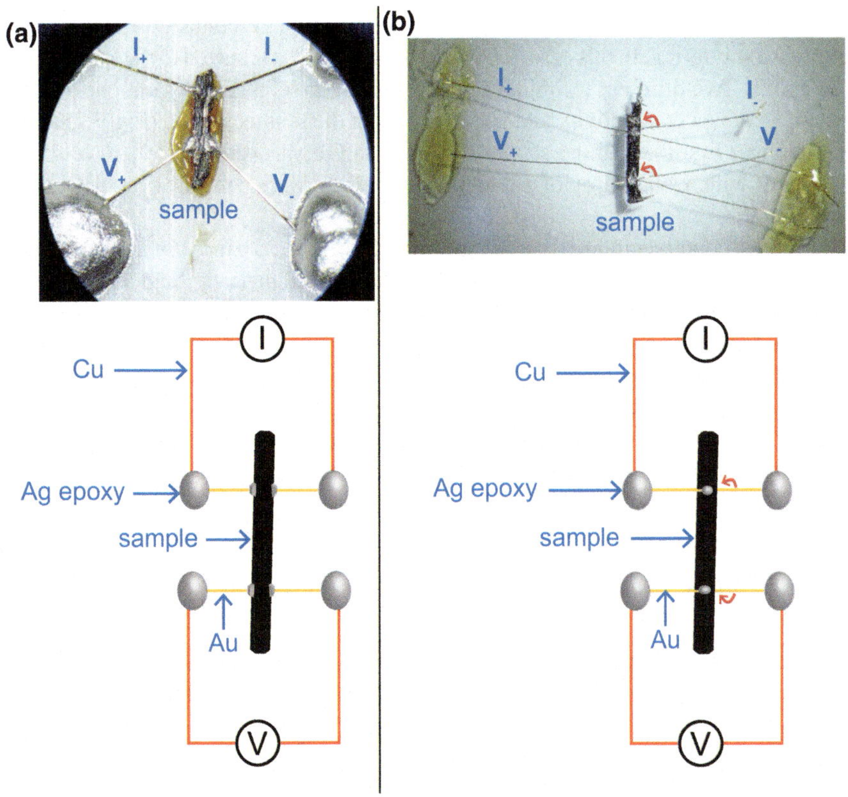

Fig. 3.28 4-Probe transversal contacts on a $Na_{2-\delta}Mo_6Se_6$ sample, using two different preparation techniques. Silver (Ag) epoxy, gold (Au) and copper (Cu) wires are indicated. **a** The four contacts are deposited on the sides of the sample, which is fixed with GE varnish. **b** The four contacts are fixed on top and bottom of the sample in two steps. The red arrows indicate the wires going under the sample. For both techniques in **a** and **b**, a schematic shows the 4-probe setup for the measurement

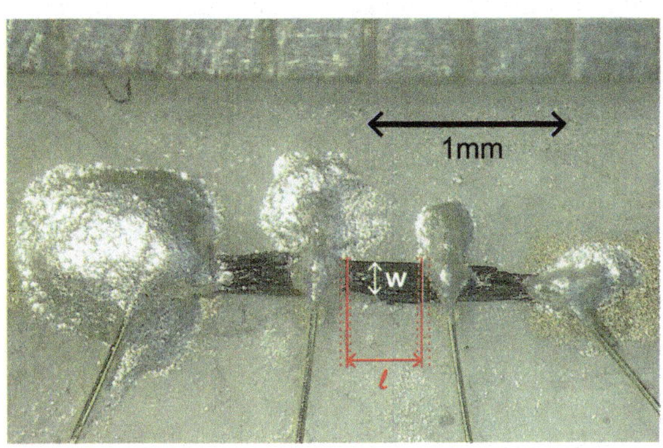

Fig. 3.29 Example of the size measurement method. The distance between the voltage contacts l is measured with the errors indicated by dashed lines. The cross section A is estimated from the sample width w and shape. Here, $l =$ 0.4 mm and $w = 0.12$ mm

In the transversal configuration, the distance between the contact was too small to allow Au sputtering. In this case, the wires were directly connected to the sample with Ag epoxy. Two different techniques were used for the contact preparation to obtain optimal contact conductivity and surface area. In the first technique (Fig. 3.28a), the sample was first fixed to the sapphire plate with GE varnish, then 50 μm or 25 μm Au wires with a small drop of Ag epoxy on their tip were deposited on the sides of the sample.

In the second technique (Fig. 3.28b), long 50 μm or 25 μm Au wires were fixed on their end with GE varnish, the sample was then slid beneath them and a drop of Ag epoxy was deposited to connect the wire to the sample. The contacts were then cured to harden them, which subsequently allowed us to flip the sample upside-down and connect the two other contacts on the other side using the same technique.

After curing, the contacts were always verified to be Ohmic ($V = IR$) at room temperature. The typical contact resistance in the longitudinal configuration was 1–2 Ω and in the transverse configuration 1–200 Ω.

3.7.3 *Transport Measurement Details*

The sizes of the samples were measured manually using an optical microscope and the measurement program imageJ. The technique being quite approximate, each measurement was repeated at least 10 times to then be averaged. The standard deviation of each averaging was then defined as the experimental error.

The resistivity values were extracted from the transport measurement as $\rho = R \cdot A/l$, where R is the resistance, A is the cross-sectional area and l the distance between the voltage contacts (Fig. 3.29). Due to the poor depth visibility of the microscope and the fragility of the crystals, the cross-sectional area was an extremely difficult parameter to estimate. Therefore, a width-to-height ratio technique was used: the width w of the sample was measured to obtain $A_{ideal} = w^2$. The shape of the sample was then estimated by tilting the sample to obtain the real A: a sample of height equal to width would have $A = A_{ideal}$, while a very flat sample would have $A = 20\% A_{ideal}$.

Finally, the errors on the resistivity were calculated from the error on the size extraction and the measurement error on R.

References

1. http://www.janis.com/
2. http://www.cryomagnetics.com/
3. http://www.qdusa.com/
4. F. Pobell, *Matter and Methods at Low Temperatures*, 3rd edn. (Springer, Berlin, 2007)
5. G.K. White, *Experimental Techniques in Low Temperature Physics* (1987)
6. http://phynelab.org/Facility/phycosmos/

7. http://www.keithley.com/
8. http://www.thinksrs.com/
9. http://www.lakeshore.com/
10. http://hollandshielding.com/
11. M. Potel, R. Chevrel, M. Sergent, J.C. Armici, M. Decroux, Ø. Fischer, New pseudo-one-dimensional metals: M2Mo6Se6 (M = Na, In, K, Tl), M2Mo6S6 (M = K, Rb, Cs), M2Mo6Te6 (M = In, Tl). J. Solid State Chem. **35**, 286–290 (1980)
12. M. Potel, P. Gougeon, R. Chevrel, M. Sergent, Labilité des cations dans les chalcogénures ternaires de molybdène: voies d'accès à de nouvelles synthèses. Revue de chimie minérale **21**(4), 509–536 (1984)
13. D. Ansermet, A.P. Petrović, S. He, D. Chernyshov, M. Hoesch, D. Salloum, P. Gougeon, M. Potel, L. Boeri, O.K. Andersen, C. Panagopoulos, Reentrant phase coherence in superconducting nanowire composites. ACS Nano **10**(1), 515–523 (2016)
14. A.P. Petrović, D. Ansermet, D. Chernyshov, M. Hoesch, D. Salloum, P. Gougeon, M. Potel, L. Boeri, C. Panagopoulos, A disorder-enhanced quasi-one-dimensional superconductor. Nat. Commun. **7**, 12262 (2016)

Chapter 4
The Quasi-One-Dimensional $Na_{2-\delta}Mo_6Se_6$

4.1 Introduction

In 1971, a new type of superconductor was discovered by R. Chevrel et al. [1]: the octahedral $A_x Mo_6 X_8$ clusters, where X = chalcogens and A_x = metals interstitial atoms with $0 < x < 4$. The crystalline structure of the compound with A_x = Pb and X = S is presented in Fig. 4.1. These Chevrel phases were shown to have relatively high values for T_c and a very high H_{c2} (15 K and 60 T in $PbMo_6S_8$, respectively [2, 3]). At the time, these materials were considered high critical temperature superconductors (high-T_c cuprates had not yet been discovered). Chevrel compounds exhibit unconventional superconducting properties of great interest from a fundamental point of view. For example, the presence of rare earth elements induces ferromagnetism in the material and has been shown to coexist with superconductivity. In $HoMo_6S_8$, a non-uniform ferromagnetic phase develops in the superconducting state and is then replaced by long-range magnetic ordering. Therefore, $HoMo_6S_8$ is only superconducting between 2 K and 0.65 K, whereas below 0.65 K the material is ferromagnetic [4].

The Chevrel clusters stimulated the development of several new Mo-based materials assembled from Mo_6X_8 clusters. The scientific enthusiasm for these materials led M. Potel (in collaboration with the discoverers of the Chevrel phases) to synthesize the q1D molybdenum chalcogenide family [5]: $M_{2-\delta}Mo_6X_6$ where M = metal, X = chalcogens and δ represents M vacancies. These crystals can be seen as linear condensations of the Chevrel building units Mo_6X_8. They are prepared by high temperature solid-state reaction (Sect. 3.3.1) and are highly stable under ambient environment, making them suitable for exploratory measurements. However, the fabrication technique leads to non stoichiometric compounds ($\delta > 0$), hence requiring careful characterization.

D. Ansermet, *Emergent Superconductivity in Low Dimensions*, Springer Theses,
https://doi.org/10.1007/978-981-13-2941-8_4

Fig. 4.1 Crystal structure of the Chevrel phase $PbMo_6S_8$. In this material, each Pb atom is surrounded by eight Mo_6S_8 units. Here only four units are shown. From Mourachkine [4]

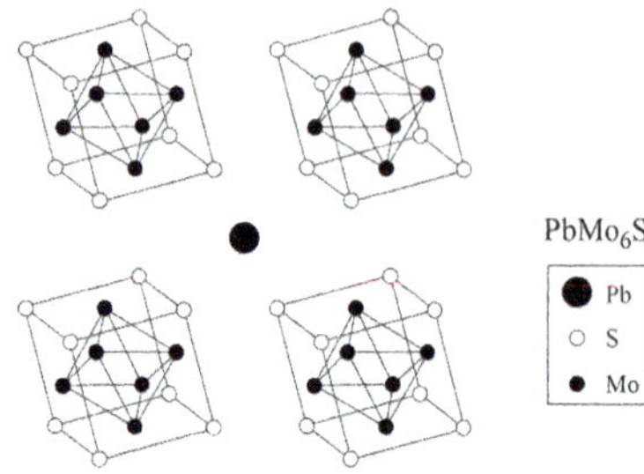

4.2 The $M_{2-\delta}Mo_6Se_6$ Family

The $M_{2-\delta}Mo_6Se_6$ compounds (with M = In, Tl, Na, K, Rb, Cs) have a highly anisotropic crystalline structure, where 1D chains of hexagonal $(Mo_6Se_6)_\infty$ unit cells are weakly coupled by M ions. The crystal structure of the compound with M = Na can be seen in Fig. 4.2. The structure is identical for all $M_{2-\delta}Mo_6Se_6$ compounds: only the radius of the M atom changes.

Mo_6Se_6 single wires (separated by dissolution of $Li_2Mo_6Se_6$ crystals) [8] exhibit a metallic ground state down to 5 K: no variations in their lattice constant or periodicity and no evidence for an energy gap opening has been reported. However, when M ions are inserted in the structure, the nature of the ground state drastically changes. A short overview of the literature [9–18] informs us that compounds with elements from group III (M = Tl, In) have a superconducting ground state, while the elements from group IA (M = Na, K, Rb, Cs) lead to an insulating ground state. For all compounds except $Na_{2-\delta}Mo_6Se_6$, these claims are confirmed by longitudinal transport measurements (Fig. 4.3). The $Na_{2-\delta}Mo_6Se_6$ compound is the major focus

Fig. 4.2 Crystal structure of $Na_{2-\delta}Mo_6Se_6$ view perpendicular (left) and parallel (right) to the chains [6, 7]

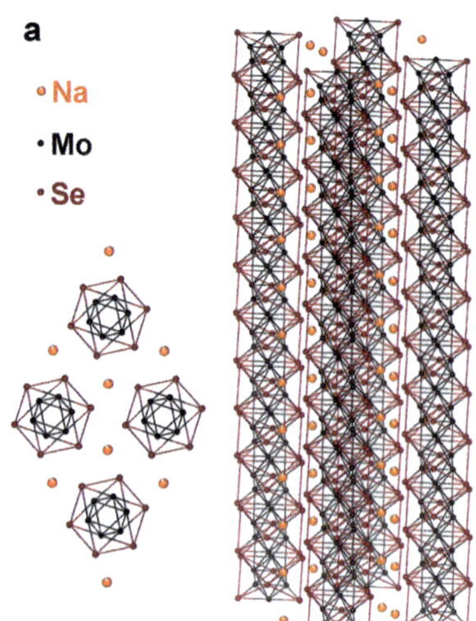

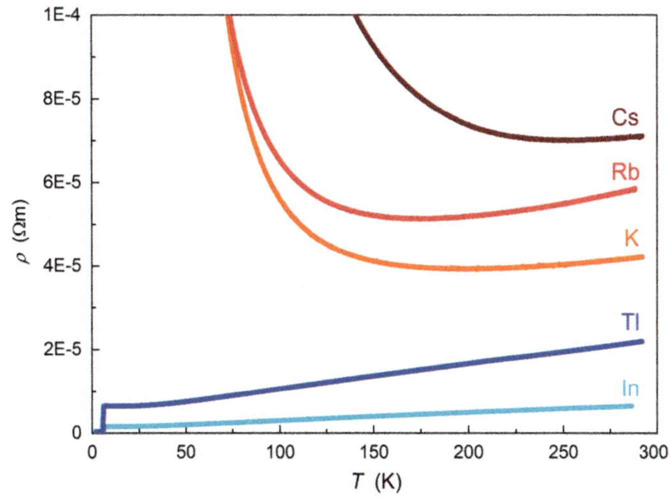

Fig. 4.3 Resistivity ρ as a function of temperature T for the different members of the $M_{2-\delta}Mo_6Se_6$ family, where M = (Tl, In, K, Rb, Cs). Data acquired in the VTI cryostat. The transport behavior of the compound with M = Na will be the focus of Chaps. 5, 6 and 7

of this thesis: while an insulating behavior is observed with decreasing temperature, a superconducting ground state emerges from the insulating instability [6, 7].

4.2.1 The Insulating Compounds

Very little data exists in the literature for $K_{2-\delta}Mo_6Se_6$, $Rb_{2-\delta}Mo_6Se_6$ and $Cs_{2-\delta}Mo_6Se_6$. They have generally been observed to undergo a gradual metal-insulator (MI) transition with decreasing temperature, leading to an insulating ground state at low temperature [10–12]. The transition was always observed to be smooth and continuous (Fig. 4.3).

The insulating ground state of these materials is believed to be inherent from the large electropositivity of the group IA elements whose orbitals are relatively higher in energy than the highest occupied levels of the Mo_6Se_6 chains. It is therefore expected that the IA metals contribute very little to the electron density at the Fermi level. The Fermi planes are barely warped and the electronic properties are believed to be more 1D than for the $Tl_{2-\delta}Mo_6Se_6$ or $In_{2-\delta}Mo_6Se_6$ compounds.

Numerous mechanisms can cause a MI transition. In the $M_{2-\delta}Mo_6Se_6$ literature, periodic modulations in the density of electronic spins or charges with a characteristic spatial frequency were mainly considered. These consist of a low temperature spin or charge-ordered state, respectively, opening an energy gap at the Fermi surface. Spin density waves (SDW) usually imply an anti-ferromagnetic ordering and can be detected by magnetic measurements. In charge density waves (CDW), the opening of an energy gap is observable in the electronic structure (by transport measurements, for example). Especially in a 1D system, CDW can take place via a structural distortion, known as the Peierls transition [19]. Quoting Peierls, in a crystalline structure, *a one-dimensional equally spaced chain with one electron per ion is unstable*. This can be explained as follows: in a 1D crystal with lattice spacing a, the lattice periodicity leads to energy band gaps in the $E - \mathbf{k}$ diagram at $\mathbf{k} = n\pi/a$ (n=integer). For a half-filled band, each ion contributes one electron, up to $\mathbf{k} = \pm\pi/2a$. If lattice distortion takes place such as every other ion moves towards each other, the lattice periodicity doubles to $2a$, which effectively introduces new band gaps with $\mathbf{k} = n\pi/2a$. This causes small energy savings, based on the distortion of the bands in the vicinity of new gaps: the electrons stay at a lower energy than they would be in the original lattice. Therefore, this lattice distortion becomes energetically favorable once this energy saving is larger than the elastic energy cost of the ion rearrangement.

Experimentally, Peierls transitions can be detected by changes in the crystal structure with lowering temperature.

A few works in the 1980s and 1990s attempted to explain the MI transitions observed in the $M_{2-\delta}Mo_6Se_6$ family [9, 11, 12, 20]. These studies took into account early band-structure calculations predicting 3 contributions to the Fermi surfaces (one Mo d band and two 3D electron pockets at the zone boundary). These calculations suggested that a possible structural distortion (Peierls transition) would be absent, as 3D pockets are expected to sufficiently stabilize the structure. However, these early calculations were obtained by non-self-consistent approaches, and were proved later [17] to be incorrect by a more accurate self-consistent method. The correct band structures are presented in this thesis (Sect. 4.4): only a very flat, single Mo d_{xy} band crosses the Fermi level [6, 7, 17]. Structural distortions could therefore take place in this system.

From the recent literature on these insulating materials, the presence of a SDW was refuted in $Rb_{2-\delta}Mo_6Se_6$ [17], but no clear proof supports the presence or absence of a CDW mechanism. Therefore, understanding the mechanism driving this MI transition remains an open problem.

4.2.2 The Superconducting Compounds

$In_{2-\delta}Mo_6Se_6$ and $Tl_{2-\delta}Mo_6Se_6$ undergo a superconducting transition at $T < 10$ K (Fig. 4.3) and are the most anisotropic superconductors according to the electronic anisotropy calculated in Sect. 4.4 (Table 4.1). The M = Tl compound in particular has been studied in various works since 1980 [10, 13–18, 21–25].

The longitudinal transport behavior of $Tl_{2-\delta}Mo_6Se_6$ led to the classification of two types of samples [21]: A-type samples exposed metallic behavior down to the superconducting transition temperature, while B-type samples showed a broad minimum in the normal state resistivity around $T \sim 20-80$ K. Interpretations of the B-type behavior suggested a CDW origin [17].

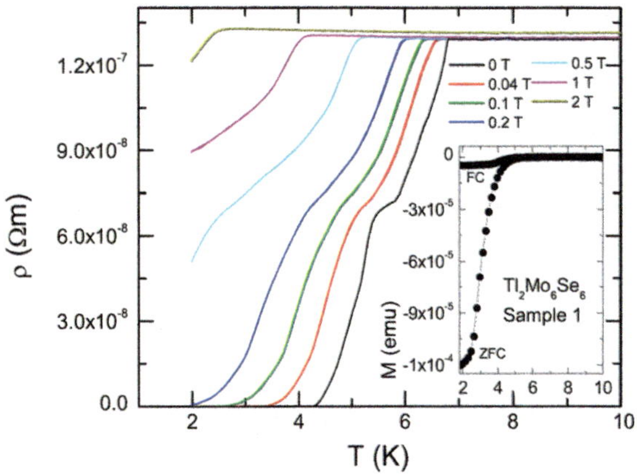

Fig. 4.4 Low temperature transport behaviors of $Tl_{2-\delta}Mo_6Se_6$ with magnetic field applied perpendicular to the z-axis. **Inset**: Magnetization measurement showing the emergence of a Meissner state at $T \sim 4.5$ K. Data from Bergk et al. [18]

The superconducting transition in $Tl_{2-\delta}Mo_6Se_6$ exhibits a hump, suggesting a two-step process for the establishment of superconducting coherence [18] (Fig. 4.4). Bergk et al. [18] explained this hump as a signature of a Berezinskii–Kosterlitz–Thouless (BKT) transition [26, 27], characterizing a thermally-driven transition from bound vortex-antivortex pairs to free phase vortices with increasing temperature. Additionally, a Meissner state develops in the material at temperature below the observed BKT transition (inset Fig. 4.4) and confirms phase-coherent superconductivity in $Tl_{2-\delta}Mo_6Se_6$.

$Tl_{2-\delta}Mo_6Se_6$ and $In_{2-\delta}Mo_6Se_6$ exhibit a s-wave pairing symmetry [17] and are therefore spin-singlet superconductors. Their superconducting coherence lengths are reported in Table 4.1 and their penetration depths are $\lambda_{\|} \sim 0.12, 0.13$ and $\lambda_{\perp} \sim 1.5, 2.2$ µm, respectively [15, 17].

4.2.3 *The Intermediate Compound* $Na_{2-\delta}Mo_6Se_6$

The only experimental study of $Na_{2-\delta}Mo_6Se_6$ reported in the literature was performed on powder samples [12]: the material was shown to undergo a MI transition suppressed monotonically by pressure, with an emergent superconducting state at $\sim$50 kbar. This result has never been reproduced since its publication in 1985. In this thesis it is demonstrated that $Na_{2-\delta}Mo_6Se_6$ lies at the border between superconducting and insulating instabilities, making it the most q1D superconducting material ever studied [6, 7].

The characterization and analysis of the $M_{2-\delta}Mo_6Se_6$ compounds through transport measurements is highly non-trivial. While large property variations from one compound to another are expected, important variations also occur between different samples of the same compound due to the presence of disorder. Each M ion provides 1 electron, effectively doping the 1D chains. Therefore, any M vacancy ($\delta > 0$) reduces the Fermi energy and creates intrinsic disorder. In consequence, larger δ equals larger disorder. This disorder has a double effect: first, a reduction of the coupling between the chains, and second, a deterioration in the conductivity along the chains (since M-doping contributes to the carrier density). The vacancy density in the material is directly dependent on the mobility of the M ion during crystal growth. For example, Na is much smaller than Tl and hence $Na_{2-\delta}Mo_6Se_6$ is expected to exhibit higher levels of disorder than $Tl_{2-\delta}Mo_6Se_6$ [7]. Indeed, a larger ion mobility would allow more vacancy formation during crystal growth, leading to a larger δ value.

4.3 Structural Properties

Several $Na_{2-\delta}Mo_6Se_6$ crystals were characterized by our collaborator D. Chernyshov in the European Synchrotron Radiation Facility (Grenoble, France).[1] This work provided detailed crystallographic information and X-ray diffraction (XRD) patterns to analyze the stoichiometry, nature, and distribution of the eventual disorder in the compound [6, 7].

4.3.1 *Crystal Structure*

Crystals of the $M_{2-\delta}Mo_6Se_6$ family present a highly q1D chain-like structure. They are composed of 1D $(Mo_6Se_6)_\infty$ chains, which are stacks of double Mo_3 and Se_3 triangular units, separated by M guest ions (Fig. 4.2).

The crystals were analyzed at 293 and 20 K, cooling the crystals using a helium blower. For both temperatures, the structure obtained was a hexagonal lattice with space group $P6_3/m$ in agreement with earlier works [5, 9]. At 293 K, the lattice parameters are $a = 8.65$Å, $c = 4.49$Å (with a minimum inter-chain Mo–Mo separation of 6.4 Å), while at 20 K, $a = 8.61$Å, $c = 4.48$Å. As no deviation of the crystal structure was found between 20 and 293 K, any crystal distortion such as a Peierls transition can be ruled out from occurring in this material above 20 K [6, 7].

[1]Data was acquired by D. Chernyshov at the Swiss-Norwegian Beamlines (SNBL) of the European Synchrotron Radiation Facility (Grenoble, France) at the end station BM01A, using a PILATUS2M pixel area detector [28]. A monochromatic beam at a wavelength of 0.694 Å was slit-collimated down to a size of $100\times100\ \mu m^2$. The sample-to-detector distance and the parameters of the detector were calibrated using a LaB_6 NIST standard. The detector images were recorded by ϕ-scans in shutter-free mode with a 0.1° angular step. The data were preprocessed by the SNBL Tool Box [29], followed by the CrysAlis Pro software package (Agilent Technologies, version 171.36.24). The crystal structure was solved with SHELXS and subsequently refined with SHELX [30].

4.3.2 *Intrinsic Disorder from Na Vacancies*

As expected from early crystals analyses done by M. Potel's group in Rennes (unpublished), the $Na_{2-\delta}Mo_6Se_6$ crystals are non-stoichiometric, i.e. $\delta > 0$. XRD analyses were performed on three randomly-chosen crystals. The data indicate $\delta = 0.2 \pm 0.036$, $\delta = 0.22 \pm 0.030$ and $\delta = 0.26 \pm 0.08$. This corresponds to a Na deficiency of 10, 11, and 13%, respectively. In contrast, the structural refinements indicate that the chains are highly ordered with close to 100% occupancies at the Mo and Se sites.

Since Na vacancies are expected to be the main source of disorder in the crystals, it is important to determine their spatial distribution. The observed vacancies could exhibit long or short range order, form clusters within the structure or be entirely randomly distributed. Their spatial distribution has a direct impact on their effect on the electronic properties of the material. For example, any correlation or ordering in the vacancy positions would result in a minor reduction in the electron mobility, while electron localization would uniquely emerge from unclustered randomly-distributed disorder.

An analysis by diffuse X-ray scattering was performed at room temperature on the crystal exhibiting the largest Na vacancy concentration, i.e. $\delta = 0.26$ (Fig. 4.5). Thermal diffuse scattering (TDS) is observed, giving rise to a blurring in the Bragg spots (i.e. lattice diffraction peaks). As the TDS scales with $\mathbf{Q}\langle u\rangle$, it increases with the reciprocal lattice vector $\mathbf{Q}$, where u is the average atomic displacement due to thermal lattice vibrations. The TDS intensity is controlled by structural factors and hence peaks near stronger Bragg reflections: TDS intensity also increases at higher temperature.

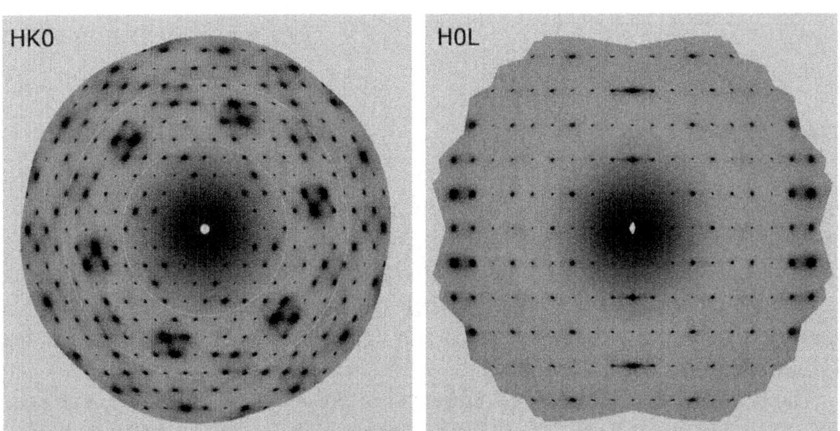

Fig. 4.5 Reconstructed layer diffraction patterns from images acquired at 300 K in the $(hk0)$ (left) and $(h0k)$ (right) planes in a $Na_{2-\delta}Mo_6Se_6$ crystal with $\delta = 0.26$. Measurements were performed by D. Chernyshov at the European Synchrotron Radiation Facility (Grenoble, France)

Various scenarios for the vacancy distribution can be considered.

Firstly, if the Na vacancies exhibited long range order, this would be visible through a set of new Bragg reflections. This is not the case here, as all the reflection points correspond to the conventional $M_{2-\delta}Mo_6Se_6$ structure.

Secondly, the Na vacancies could exhibit short-range order, such as stripes, planes, chess-board patterns, etc. This would lead to structured diffuse scattering, which appears as weak and diffuse lines connecting the diffraction spots: this is also absent from the images.

Thirdly, the vacancies could form clusters and create Huang scattering [31] due to the elastic distortion of the elastic moduli. Huang scattering would appear as clouds of diffuse scattering near the Bragg spots as distortions occurring in the crystal are caused by the size mismatch from the vacancies. This is not observed in the data.

Since only TDS is detected in the diffraction images, it is safe to conclude that disorder due to Na vacancies is randomly distributed [6, 7]. The lack of spatial correlation in the Na vacancy distribution is attributed to the elevated crystal growth temperature, as Na ions have a high mobility due to their small size and low mass.

From this study, it can be concluded that the $Na_{2-\delta}Mo_6Se_6$ crystals exhibit intrinsic disorder emerging from Na vacancies which are randomly distributed in the lattice. Since each Na atom donates 1 electron to the conduction band, any absence of Na atoms introduces an intrinsic electronic disorder into the system [6, 7].

4.4 Electronic Properties

Electronic structure calculations of $Na_{2-\delta}Mo_6Se_6$, $Rb_{2-\delta}Mo_6Se_6$, and $Tl_{2-\delta}Mo_6Se_6$ have been performed by our collaborators L. Boeri (TU Graz, Austria) and O. K. Andersen (Max Planck Institute Stuttgart, Germany).

4.4.1 Band Structure

Our collaborators performed ab initio density functional theory (DFT) calculations of the electronic structure of stoichiometric $Na_{2-\delta}Mo_6Se_6$ ($\delta = 0$), using the internal coordinates obtained from the X-ray structural refinement and the full-potential linear augmented-plane-wave method [32]. Their calculations provided nearly identical features to those previously published [17] for $Rb_{2-\delta}Mo_6Se_6$, $In_{2-\delta}Mo_6Se_6$, and $Tl_{2-\delta}Mo_6Se_6$ [6, 7].

The complete band structure is presented in Fig. 4.6 for $\mathbf{k}$ along the central $\mathbf{k}_z$-axis (ΓA) and along lines (AL-LH-HA) perpendicular to it on the Brillouin-zone boundary ($\mathbf{k}_z = \pi/c$), with a zoom on the area near the Fermi level E_F (Fig. 4.7). The z-axis refers to the reciprocal chain axis while x is normal to the chains. A 3D cartoon of the BZ boundary is presented in Fig. 4.8 together with the real space crystal structure.

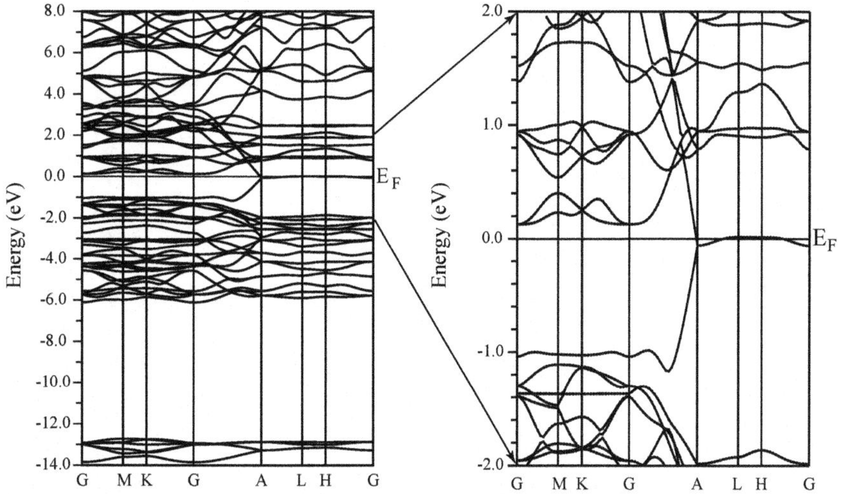

Fig. 4.6 Electronic structure of $Na_{2-\delta}Mo_6Se_6$ ($\delta = 0$) calculated using DFT. On the right the single Mo d_{xz} band crossing the Fermi level E_F is highlighted. Calculations were performed by L. Boeri and O. K. Andersen

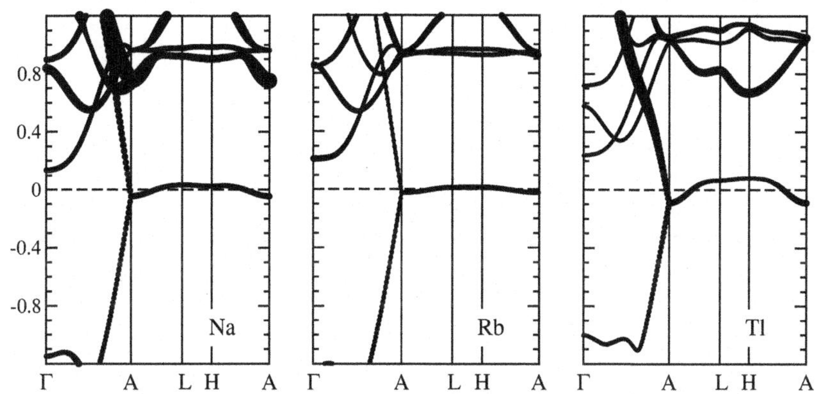

Fig. 4.7 Calculated band structures of $M_{2-\delta}Mo_6Se_6$ ($\delta = 0$) for M = Na, Rb, and Tl decorated with M valence-orbital characters. Energies are in eV with respect to the Fermi level of the stoichiometric compounds. Calculations were performed by L. Boeri and O.K. Andersen

The band structures exhibit a single spin-degenerate band of Mo d_{xz} character which crosses E_F at the Brillouin zone (BZ) boundary. Importantly, the band is not gapped at the BZ boundary: a translation by $c/2$ followed by 180°-rotation around any z-axis covers all possible combinations of the crystal. All representations of this Abelian group are enumerated by choosing $\mathbf{k}$: the band structure can be folded out to a BZ twice as high with a boundary at $\mathbf{k}_z = \pi/(c/2)$ and a formula $M Mo_3Se_3$ [33].

Parallel to the $(Mo_6Se_6)_\infty$ chains (in the longitudinal direction), the conduction band is highly dispersive, which suggests a high conductivity. On the other hand, it

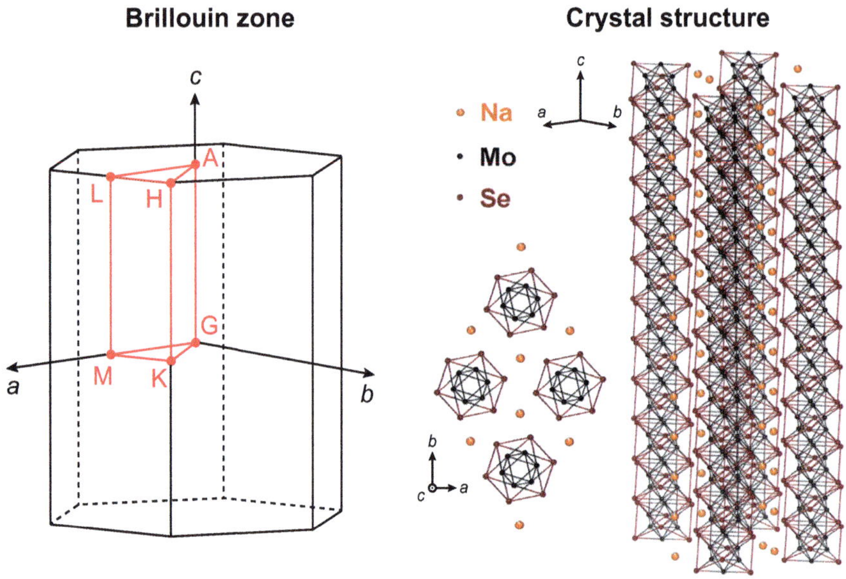

Fig. 4.8 Brillouin zone boundary for a hexagonal lattice structure. The letters in red correspond to the lattice points where the electronic band structures are calculated. The real space crystal structure of $Na_{2-\delta}Mo_6Se_6$ is also presented to highlight the high-conductive direction (along the c-axis)

is weakly dispersive perpendicular to the chains (in the transversal direction). The electronic band structure highlights the large structural anisotropy in $Na_{2-\delta}Mo_6Se_6$.

The dispersion of the conduction band can be characterized by the electron hopping integrals, which describe the electron wavefunction overlap on neighboring sites within a tight-binding scenario and hence control the conductivity. The intra-chain hopping energy $t_{\parallel}$ represents the longitudinal electron flow, while the inter-chain hopping energy $t_{\perp}$ corresponds to the transverse transport. The band structures have been calculated including the M-valence orbital characters in the $Na_{2-\delta}Mo_6Se_6$, $Rb_{2-\delta}Mo_6Se_6$ and $Tl_{2-\delta}Mo_6Se_6$ compounds (Fig. 4.7). These are used to extract numerical values of $t_{\parallel}$ and $t_{\perp}$ in the different compounds.

The value of the longitudinal hopping integral is obtained as follows: the conduction band can be seen to have a width W from a linear extrapolation and folding out to distance $2\Gamma A = \pi/(c/2)$. The value of W is 7.4 eV for $Na_{2-\delta}Mo_6Se_6$ which corresponds closely to the values for M = K, Rb, In and Tl [17]. A corresponding velocity component is obtained from $v_{\parallel} = Wc/(2\pi)$ and, taking the dispersion to be $-2t_{\parallel}\cos(k_z c/2)$, the hopping integral is calculated as $t_{\parallel} = W/(2\pi) = 1.2$ eV for $Na_{2-\delta}Mo_6Se_6$. This value is nearly the same in the other compounds, as seen in Table 4.1. Hopping along the chain takes place between neighboring Mo_3-molecular orbitals: each of these is the anti-bonding linear combination of the 3 atomic $4d_{xz}$ orbitals on the Mo_3-triangle [6, 7].

Table 4.1 Electronic anisotropy in q1D materials

	$t_\parallel$		$t_\perp$		$t_\parallel/t_\perp$	$t_\perp^2/t_\parallel$
	[eV]	[K]	[meV]	[K]	[–]	[K]
$Tl_{2-\delta}Mo_6Se_6$	1.03	12000	20	230	52	4.4
$In_{2-\delta}Mo_6Se_6$	1.03	12000	16	190	63	3.0
$Na_{2-\delta}Mo_6Se_6$	1.20	14000	10	120	117	1.0
$K_{2-\delta}Mo_6Se_6$	1.20	14000	5	60	233	0.3
$Rb_{2-\delta}Mo_6Se_6$	1.20	14000	3	30	467	0.06
$(TMTSF)_2ClO_4$ [36]	0.26	3000	26	300	10	30
$(TMTSF)_2PF_6$ [37]	0.26	3000	20	230	13	18
$LiMo_6O_{17}$ [38]	0.80	9300	15	170	55	3.3
$LiMo_6O_{17}$ [39]	0.74	8600	20	230	37	6.2

To calculate the transversal hopping integral, the in-plane linewidth w of the conduction band is considered. In $Na_{2-\delta}Mo_6Se_6$, $Rb_{2-\delta}Mo_6Se_6$ and $Tl_{2-\delta}Mo_6Se_6$, $w = 90$, 23, and 180 meV, respectively. The relative energies at A, L, and H are respectively $-6t_\perp$, $2t_\perp$, and $3t_\perp$ in terms of the hopping integral between neighboring chains. This yields $t_\perp = w/9 = 10$, 3, and 20 meV for $Na_{2-\delta}Mo_6Se_6$, $Rb_{2-\delta}Mo_6Se_6$, and $Tl_{2-\delta}Mo_6Se_6$, respectively. Here, $t_\perp$ is strongly dependent on the nature of the M ion. Hopping depends on the inter-chain distance (which is larger for Rb than for Na for example) and proceeds via the M cation [6, 7].

The obtained values are summarized in Table 4.1, together with values of a few other well known q1D superconductors such as the Bechagaard salts [34] and purple bronze [35] for comparison. From the values of $t_\perp$, it is clear that the superconducting compounds of the $M_{2-\delta}Mo_6Se_6$ family are the most q1D superconductors known to date [6, 7].

Now, let us turn to the calculations of the density of states (DoS).

The band dispersion is expressed using a tight-binding expression [40] and the global cartesian xyz-system with y pointing between nearest - neighbor chains:

$$E(\mathbf{k}) = -2t_\parallel \cos\left(\frac{c}{2}k_z\right) - 2t_\perp \left\{\cos\left(ak_y\right) + \cos\left(\frac{a}{2}\left(k_y + \sqrt{3}k_x\right)\right) + \cos\left(\frac{a}{2}\left(k_y - \sqrt{3}k_x\right)\right)\right\} \tag{4.1}$$

In order to calculate the DoS $N(E)$, it is necessary to consider its stability towards Na vacancies. With $w << W$ being so small, the DoS is simply $N(E) = 1/W$ states per spin per MMo_3Se_3 independently of w and E, as long as the dispersion is linear and no other band conducts. It is also independent of the M-stochiometry, from approximately $M_{0.75}$–$M_{1.05}$. From the XRD pattern, the measured Na content of the crystals ($\delta \leq 0.2$) lies within this range. $N(E) = 0.135$ and 0.166 states/(eV×spin×MMo_3Se_3) are obtained for $Na_{2-\delta}Mo_6Se_6$ and $Tl_{2-\delta}Mo_6Se_6$, respectively. These values correspond to a bare DoS: they have to be renormalized

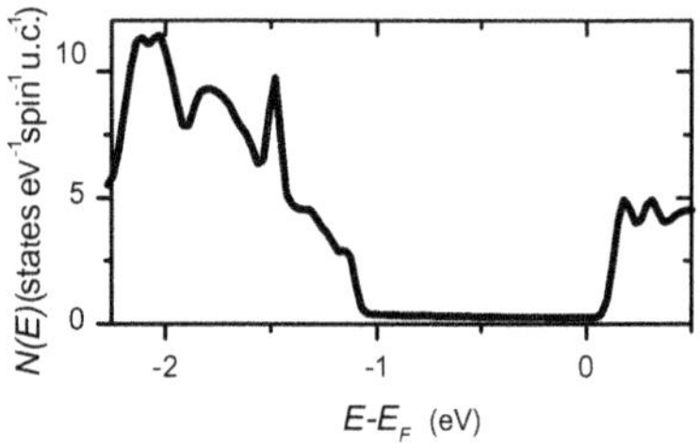

Fig. 4.9 Electronic density of states $N(E)$ close to the Fermi level in $Na_{2-\delta}Mo_6Se_6$. $N(E_F)$ remains constant in $Na_{2-\delta}Mo_6Se_6$ for $-0.1 < \delta < 0.5$. Calculations were performed by L. Boeri

by electron–phonon (e–ph) and e–e interactions with a mass-enhancement factor $1 + \lambda$. Renormalization was performed using the ratio between the experimental electronic specific heat coefficient γ and the bare DoS. However, the values of γ are known only for M = In, Tl and Rb [17]. An identical mass enhancement is assumed between $Na_{2-\delta}Mo_6Se_6$ and $In_{2-\delta}Mo_6Se_6$, to finally obtain an enhanced DoS of 0.17 states/(eV×spin×MMo$_3$Se$_3$) for $Na_{2-\delta}Mo_6Se_6$ [6, 7]. The calculated density of states $N(E)$ is presented in Fig. 4.9 as a function of $E - E_F$.

The DoS remains extremely low until $E - E_F \sim -1$ eV. The visible peak appearing in $N(E)$ below $E - E_F \sim -1$ eV emerges due to a reduced dispersion in the conduction band, together with a multitude of additional bands at lower energy (bands below –1 eV in Fig. 4.6). Peaks in the DoS situated relatively close to the Fermi energy imply large values of the background dielectric function and hence an enhanced screening of the Coulomb repulsion.

4.4.2 Anisotropic Evolution in the $M_{2-\delta}Mo_6Se_6$ Family

The different members of the $M_{2-\delta}Mo_6Se_6$ family can be compared by plotting $t_\perp$ as a function of the M atomic radius (Fig. 4.10). As expected from the calculations, there is a clear correlation between $t_\perp$ and the atomic radius [7]. $t_\perp$ is therefore a key parameter to differentiate the members of the $M_{2-\delta}Mo_6Se_6$ family: as $t_\perp$ is reduced, the system becomes more 1D and the electronic ground state changes. $Tl_{2-\delta}Mo_6Se_6$ and $In_{2-\delta}Mo_6Se_6$ are known to become superconducting, while $K_{2-\delta}Mo_6Se_6$ and $Rb_{2-\delta}Mo_6Se_6$ become insulating. $Na_{2-\delta}Mo_6Se_6$ sits at the border between insulating and superconducting instabilities.

The value of $t_\perp = 120$ K for $Na_{2-\delta}Mo_6Se_6$ is also experimentally interesting, suggesting that within the temperature range of the measurement systems available (Chap. 3), it may be possible to track the dimensional crossover between 1D and anisotropic 3D transport.

Anisotropy can be quantified using different methods. From the definitions of the hopping energies, $t_\parallel/t_\perp$ is a useful tool to compare filamentary systems. The values are listed in Table 4.1. Alternatively, one can experimentally measure the longitudinal and transversal electrical conductivities, $\sigma_\parallel$ and $\sigma_\perp$. Their ratio

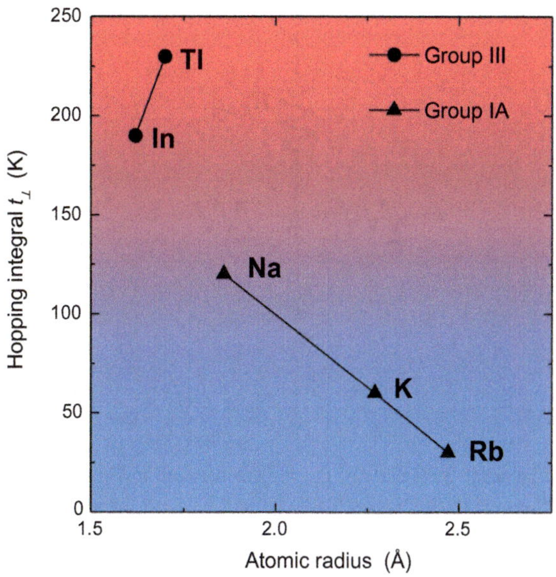

Fig. 4.10 Electronic anisotropy in the different $M_{2-\delta}Mo_6Se_6$ compounds. The colors represent the nature of the ground state: red is superconducting while blue is insulating

$\sigma_\parallel/\sigma_\perp$ gives the measured transport anisotropy. In $Na_{2-\delta}Mo_6Se_6$, $t_\parallel/t_\perp = 117$ and $\sigma_\parallel/\sigma_\perp = 3200$, while in $Tl_{2-\delta}Mo_6Se_6$, $t_\parallel/t_\perp = 52$ and $\sigma_\parallel/\sigma_\perp = 1000$. For both measures, the anisotropy at least doubles from $Tl_{2-\delta}Mo_6Se_6$ to $Na_{2-\delta}Mo_6Se_6$.

The high anisotropy emerging from the electronic structure is exacerbated by the presence of disorder ($\delta > 0$). To understand its effect on the electron transport, a simulation tool was developed to study the electron path in a large anisotropic medium [6].

4.4.3 Anisotropic Random Resistor Network

The presence of vacancies leads to an inhomogeneous electron flow in the material, as the electrons have to take a percolative route to overcome disorder. This inhomogeneous electron flow can be represented by simulating an anisotropic random resistor network for q1D materials such as the $M_{2-\delta}Mo_6Se_6$ compounds [6].

The model consists of a 2D $m \times n$ array of nodes, each connected to its 4 nearest neighbors by a resistor (Fig. 4.11). The anisotropy is simulated by setting the values of transverse resistance R_{trans} and the longitudinal resistance R_{longi} as $R_{trans}/R_{longi} = 1000$. This value was chosen to represent the order of magnitude of the measured $\sigma_\parallel/\sigma_\perp$ in $Na_{2-\delta}Mo_6Se_6$ and $Tl_{2-\delta}Mo_6Se_6$. Therefore, as expected in a real crystal, a current injected at the base of the array principally flows vertically through R_{longi}. In addition, the presence of disorder is simulated by randomly increasing 10% of R_{longi} by a factor of 10^9, simulating 10% M ion vacancies forcing the electrons to take a percolative path.

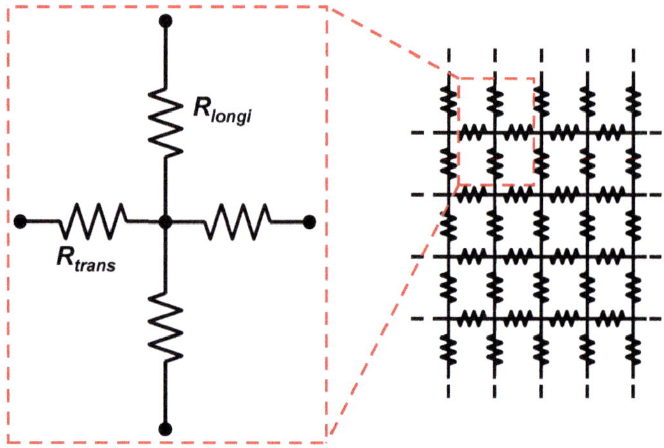

Fig. 4.11 Schematic of the resistor network to simulate electrical flow through a q1D material. R_{trans} is the transverse resistance and R_{longi} is the longitudinal resistance

The current distribution is calculated by applying Kirchoff's law to each node:

$$\sum_j \sigma_{ij}(V_i - V_j) = 0 \tag{4.2}$$

where $\sigma_{ij} \equiv 1/R_{ij}$ are the conductances between nodes i and j and V is the voltage. Boundary conditions are chosen such that $V = 1$ at the base of the array and $V = 0$ at the top. This creates a set of $m \times n$ coupled simultaneous equations, which are solved by matrix inversion.

The simulations for a 120 × 120 array are shown in Fig. 4.12. For randomly distributed vacancies (which is the case for $Na_{2-\delta}Mo_6Se_6$ as confirmed by the XRD analysis), this leads to a highly inhomogeneous current flow [6]. To analyze in more detail the electrical flow (Fig. 4.12b), the simulation is evaluated at three different areas by tracing linecuts perpendicular to the current (yellow lines). The result is presented in Fig. 4.12c: it shows that roughly 50% of the current is carried by only ~10% of the filaments. This confirms that the electrical transport is highly anisotropic across macroscopic disordered $Na_{2-\delta}Mo_6Se_6$ crystals [6].

As $Tl_{2-\delta}Mo_6Se_6$ is the most anisotropic superconductor known until know, this suggests that $Na_{2-\delta}Mo_6Se_6$ is an even more extreme q1D material to study due to its crystalline structure and the large presence of Na vacancies. In the following chapters, it will be shown that $Na_{2-\delta}Mo_6Se_6$ exhibits a superconducting ground state emerging from a strongly localized electron state, unveiling the emergence of fascinating physical properties in low dimensions.

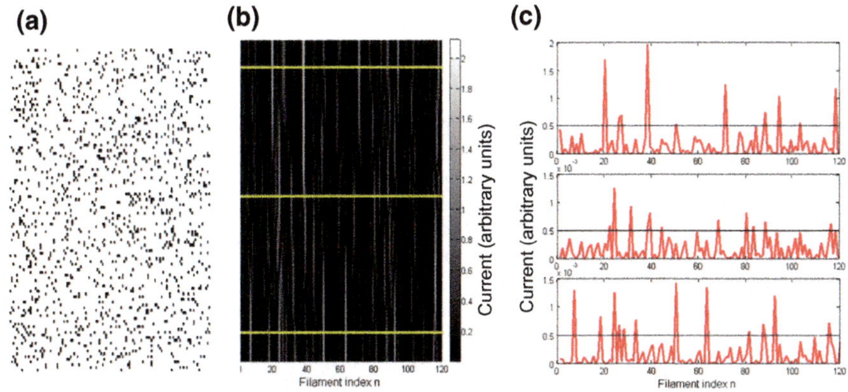

Fig. 4.12 Simulation of highly percolative current flow in a disordered q1D material. **a** Typical random vacancy distribution in a 120 × 120 array. **b** Results of the simulation of the current distribution, calculated using Eq. 4.2. The yellow lines indicate the location where the current distribution is evaluated. **c** Results of the current distribution evaluation. Simulations were performed by A.P. Petrović

References

1. R. Chevrel, M. Sergent, J. Prigent, Sur de nouvelles phases sulfurées ternaires du molybdène. J. Solid State Chem. **3**(4), 515–519 (1971)
2. R. Odermatt, Ø. Fischer, H. Jones, G. Bongi, Upper critical fields of some ternary molybdenum sulphides. J. Phys. C: Solid State Phys. **7**, L13–L15 (1974)
3. Ø. Fischer, H. Jones, G. Bongi, M. Sergent, R. Chevrel, Measurements of critical fields up to 500 kG in the ternary molybdenum sulphides. J. Phys. C: Solid State Phys. **7**, L450–L453 (1974)
4. A. Mourachkine, *Room-Temperature Superconductivity* (Cambridge International Science Publishing, Cambridge, 2006)
5. M. Potel, R. Chevrel, M. Sergent, J.C. Armici, M. Decroux, Ø. Fischer, New pseudo-one-dimensional metals: M2Mo6Se6 (M = Na, In, K, Tl), M2Mo6S6 (M = K, Rb, Cs), M2Mo6Te6 (M = In, Tl). J. Solid State Chem. **35**, 286–290 (1980)
6. D. Ansermet, A.P. Petrović, S. He, D. Chernyshov, M. Hoesch, D. Salloum, P. Gougeon, M. Potel, L. Boeri, O.K. Andersen, C. Panagopoulos, Reentrant phase coherence in superconducting nanowire composites. ACS Nano **10**(1), 515–523 (2016)
7. A.P. Petrović, D. Ansermet, D. Chernyshov, M. Hoesch, D. Salloum, P. Gougeon, M. Potel, L. Boeri, C. Panagopoulos, A disorder-enhanced quasi-one-dimensional superconductor. Nat. Commun. **7**, 12262 (2016)
8. L. Venkataraman, C. Lieber, Molybdenum selenide molecular wires as one-dimensional conductors. Phys. Rev. Lett. **83**(25), 5334–5337 (1999)
9. J.M. Tarascon, G.W. Hull, F.J. DiSalvo, A facile synthesis of pseudo one-monodimensional ternary molybdenum chalcogenides M2Mo6X6. Mater. Res. Bull. **19**(7), 915–924 (1984a)
10. J.M. Tarascon, F.J. DiSalvo, J.V. Waszczak, Physical properties of several M2Mo6X6 compounds. Solid State Commun. **52**(3), 227–231 (1984b)
11. P.H. Hor, W.C. Fan, L.S. Chou, R.L. Meng, C.W. Chu, J.M. Tarascon, M.K. Wu, Study of the metal-semiconductor transition in Rb2Mo6Se6, Rb2Mo6Te6 and Cs2Mo6Te6 under pressures. Solid State Commun. **55**(3), 231–235 (1985a)
12. P.H. Hor, R.L. Meng, C.W. Chu, J.M. Tarascon, M.K. Wu, High Pressure Study on quasi-one-dimensional compounds M2Mo6X6. Physica 135B **135**(1-3), 245–247 (1985)

13. R. Brusetti, P. Monceau, M. Potel, P. Gougeon, M. Sergent, The exotic superconductor Tl2Mo6Se6 investigated by low field magnetization measurements. Solid State Commun. **66**(2), 181–187 (1988)
14. R. Brusetti, A. Briggs, O. Laborde, M. Potel, P. Gougeon, Superconducting and dielectric instabilities in Tl2Mo6Se6: unusual transport properties and unsaturating critical field. Phys. Rev. B **49**(13), 8931–8943 (1994)
15. A.P. Petrović, Y. Fasano, R. Lortz, M. Decroux, M. Potel, R. Chevrel, Ø. Fischer, Unconventional resistive transitions in the extreme type-II superconductor Tl2Mo6Se6. Phys. C: Supercond. **460–462**(1), 702–703 (2007)
16. A.P. Petrović, Unconventional superconductivity and strong electronic correlations in molybdenum cluster compounds. Ph.D. thesis, Departement de Physique de la Matiere Condensee, faculte des Sciences, University of Geneva (2009)
17. A.P. Petrović, R. Lortz, G. Santi, M. Decroux, Phonon mode spectroscopy, electron-phonon coupling, and the metal-insulator transition in quasi-one-dimensional M2Mo6Se6. Phys. Rev. B **82**(23), 235128 (2010)
18. B. Bergk, A.P. Petrović, Z. Wang, Y. Wang, D. Salloum, P. Gougeon, M. Potel, R. Lortz, Superconducting transitions of intrinsic arrays of weakly coupled one-dimensional superconducting chains: the case of the extreme quasi-1D superconductor Tl2Mo6Se6. New J. Phys. **13**(10), 103018 (2011)
19. R.E. Peierls, *More surprises in theoretical physics* (Princeton University Press, Princeton, 1991)
20. G.X. Tessema, Y.T. Tseng, M.J. Skove, E.P. Stillwell, R. Brusetti, P. Monceau, M. Potel, P. Gougeon, Stress-induced metal-to-nonmetal transition in the quasi-one-dimensional superconductor Tl2Mo6Se6. Phys. Rev. B **43**(4), 3434–3437 (1991)
21. J.C. Armici, M. Decroux, Ø. Fischer, M. Potel, R. Chevrel, M. Sergent, A new pseudo-one-dimensional superconductor: Tl2Mo6Se6. Solid State Commun. **33**(254), 607–611 (1980)
22. S.Z. Huang, J.J. Mayerle, R.L. Greene, M.K. Wu, C.W. Chu, Pressure dependence of superconductivity in the pseudo-one-dimensional compound Tl2Mo6Se6 (1983)
23. T. Mori, Y. Yokogawa, A. Kobayashi, Y. Sasaki, H. Kobayashi, Thermoelectricity of pseudo-one-dimensional ternary chalcogenides: Tl2Mo6Se6 and In2Mo6Se6. Solid State Commun. **49**(3), 249–252 (1984)
24. R. Lepetit, P. Monceau, Superconductivity of the linear chain compound TI2Mo6Se6. J. Low Temp. Phys. **56**(3/4), 219–235 (1984)
25. H.P. Geserich, G.Scheiber, M.Dürrler, M.Potel, M.Sergent, P.Monceau, Polarized reflectance spectra of the one-dimensional superconductor Tl2Mo6Se6-a comparison with the anisotropic superconductor (SN)x (1986)
26. V.L. Berezinskii, Destruction of long-range order in one-dimensional and two-dimensional systems having a continuous symmetry group I. Classical systems. Sov. Phys. JETP **32**(3), 2–9 (1971)
27. J.M. Kosterlitz, D.J. Thouless, Long range order and metastability in 2d solids ans superfluids. J. Phys. C **5**, L124–L126 (1972)
28. B. Henrich, A. Bergamaschi, C. Broennimann, R. Dinapoli, E.F. Eikenberry, I. Johnson, M. Kobas, P. Kraft, A. Mozzanica, B. Schmitt, PILATUS: A single photon counting pixel detector for X-ray applications. Nucl. Instrum. Methods Phys. Res. Sect. A: Accel. Spectrom. Detect. Assoc. Equip. **607**(1), 247–249 (2009)
29. V. Dyadkin, *SNBL Tool Box, Release 2012–1* (Swiss Norwegian Beam Lines at ESRF, Grenoble, France, 2012)
30. G.M. Sheldrick, *SHELXL97* (University of Gottingen, Germany, 1997)
31. H.H. Huang, X-ray reflexions from dilute solid solutions. Proc. R. Soc. A **190**, 1947 (1020)
32. O.K. Andersen, Linear methods in band theory. Phys. Rev. B **12**(8), 3060–3083 (1975)
33. P. Kelly, O.K. Andersen, *Superconductivity in d-and f-Band Metals* (Academic Press, New York, 1982)
34. D. Jerome, A. Mazaud, M. Ribault, K. Bechgaard, Superconductivity in a synthetic organic conductor (TMTSF)2PF6. J. de Phys. Lett. **41**(4), 95–98 (1980)

35. M. Greenblatt, W.H. McCarroll, R. Neifeld, M. Croft, J.V. Waszczak, Quasi two-dimensional electronic properties of the lithium molybdenum bronze, Li0.9Mo6O17. Solid State Commun. **51**(9), 671–674 (1984)
36. T. Giamarchi, Theoretical framework for quasi-one dimensional systems. Chem. Rev. **104**(11), 5037–5056 (2004)
37. J. Moser, M. Gabay, P.A. Senzier, D. Jérome, K. Bechgaard, J.M. Fabre, Transverse transport in (TM)2X organic conductors: possible evidence for a Luttinger liquid. Eur. Phy. J. B: Condens. Matter Phys. **1**(1), 39–46 (1998)
38. P. Chudzinski, T. Jarlborg, T. Giamarchi, Luttinger-liquid theory of purple bronze Li0.9Mo6O17 in the charge regime. Phys. Rev. B **86**(7), 075147 (2012)
39. M. Nuss, M. Aichhorn, Effective model for the electronic properties of quasi-one-dimensional purple bronze Li0.9Mo6O17 based on ab inition calculations. Phys. Rev. B **89**(4), 045125 (2014)
40. H. Nohl, W. Klose, O.K. Andersen, Ø. Fischer, M.B. Maple, *Superconductivity in Ternary Compounds*, vol. 1 (Springer, Berlin, 1982)

Chapter 5
The Electronic Normal State in $Na_{2-\delta}Mo_6Se_6$

5.1 Introduction

To study the emergence of superconductivity in $Na_{2-\delta}Mo_6Se_6$, it is first important to understand the nature of its normal state. The presence of e–e interactions will influence the emergence and stability of the superconducting state [1–10].

In $Na_{2-\delta}Mo_6Se_6$, the electronic structure calculations (Chap. 3) predict the electronic state to be 1D at $T > t_\perp$ and anisotropic 3D at $T < t_\perp$ (where $t_\perp \sim 120$ K). However, any e–e interaction renormalizes the hopping energies. The temperature range accessible in the different measurement systems is 300−1.6 K (VTI or PPMS) and 4−0.02 K (dilution refrigerator). This allows us to study the dimensional crossover in the transport data.

We know from the sample fabrication (Sect. 3.7) and structural properties (Sect. 4.3) that the $Na_{2-\delta}Mo_6Se_6$ crystals expose different values of δ, i.e. intrinsic disorder. This suggests that each crystal should present slight variations in the transport behavior. The observations and analyses presented in this chapter were published in 2016 in *Nature Communications* [11].

5.2 General Observations

For this project, a large number of crystals have been measured and characterized. Due to their fragile nature, especially under repeated large thermal cycling, most of the crystals showed a finite lifetime and broke during measurements: it was not possible to fully characterize each one of the crystals. Nevertheless, a relatively complete set of data has been obtained on six distinctive samples (Fig. 5.1). The crystals have been named with capital letters, from A to F, and their transport data are color-coded. The crystal names and color codes are systematically followed in Chaps. 5 and 6. The samples have been labeled systematically from A to F following

D. Ansermet, *Emergent Superconductivity in Low Dimensions*, Springer Theses,
https://doi.org/10.1007/978-981-13-2941-8_5

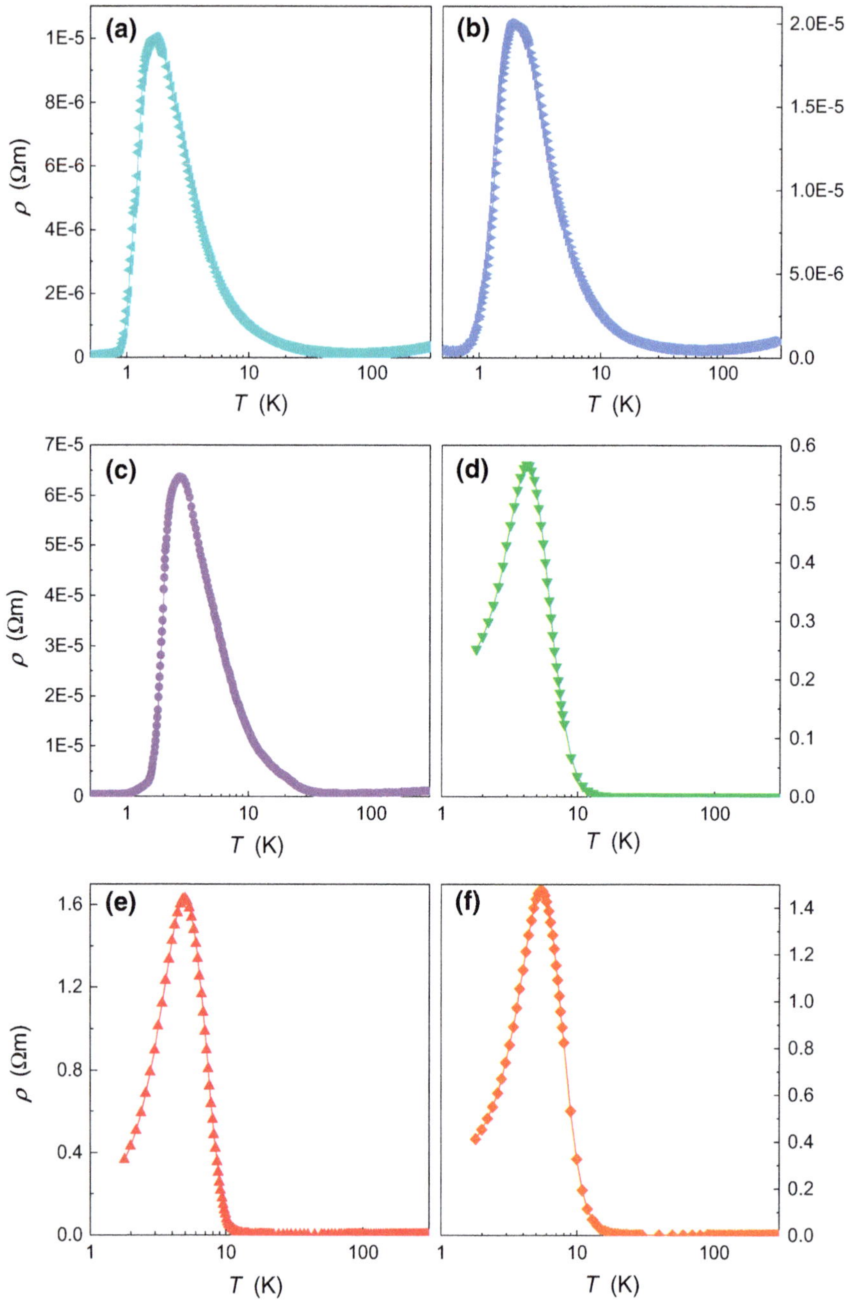

Fig. 5.1 $R(T)$ plotted on a semi-logarithmic scale for six $Na_{2-\delta}Mo_6Se_6$ crystals, where crystals A to F are labeled as figures (**a**) to (**f**), respectively. Only crystals A, B, and C were measured in the dilution refrigerator, yielding data in the mK range. All data were acquired at $I_{ac} = 10\ \mu A$ and $\omega = 1$ Hz

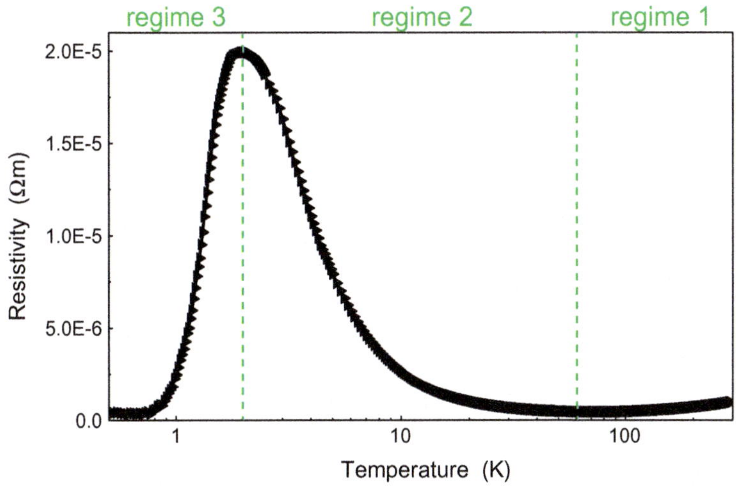

Fig. 5.2 The three different temperature regimes in the transport measurements: regime 1 where $dR/dT > 0$; regime 2 where $dR/dT < 0$; regime 3 where a superconducting instability develops. The transport data is from sample *B*

an evolutive trend observed in their transport data corresponding to a systematic increase of disorder (as we will see in Sect. 5.4).

In addition, three more crystals named in lowercase g, h, i have provided reproducible data at low temperatures ($<100-2$ K). Their high temperature transport was not obtained due to mechanical failure: these crystals were extremely small and fragile. This is the reason why the have been named using lowercase, as they are not mainly contributing to the data set and analysis. Transport data from these samples will be presented in Sects. 5.4 and 5.5.

Transport data from crystals A to F show three clear regimes as a function of temperature (Figs. 5.1 and 5.2):

1. Metallic behavior $dR/dT > 0$ at high temperature ($\sim$300 to $\sim$150 K);
2. A minimum leading to insulating behavior $dR/dT < 0$ at intermediate temperature ($\sim$150 to 2–6 K);
3. A maximum which suggests the emergence of a superconducting instability at low temperature ($<2-6$ K).

The numerical values for the main transport behaviors in each crystal are summarized in Table 5.1. The following experimental parameters were extracted: the size of the crystal corresponding to the cross-section area A divided by the length l; the value of the room temperature resistivity $\rho(300\ \text{K})$; the metal-insulator transition temperature T_{MI} corresponding to the minimum at intermediate temperature and its resistivity $\rho(T_{MI})$; the onset temperature of superconducting fluctuations T_{ons} corresponding to the maximum at low temperature and its resistivity $\rho(T_{ons})$.

Table 5.1 Characterization of the measured $Na_{2-\delta}Mo_6Se_6$ crystals

Crystal	Size A/l [m]	$\rho(300\ K)$ [Ωm]	T_{MI} [K]	$\rho(T_{MI})$ [Ωm]	T_{ons} [K]	$\rho(T_{ons})$ [Ωm]
A	1.6e-6±3e-7	3.41e-7±3.2e-8	72	1.10e-7±1.5e-8	1.68	4.1e-6±2e-7
B	3.0e-6±3e-7	8.22e-7±6.0e-8	73	3.60e-7±2.9e-8	2.01	1.7e-5±1e-6
C	6.1e-5±5e-6	1.10e-6±1.2e-7	60	4.70e-7±3.5e-8	2.75	6.4e-5±3e-6
D	7.2e-6±4e-7	1.17e-6±1.4e-7	110	8.14e-7±5.1e-8	4.20	0.3±0.02
E	3.0e-5±2e-6	1.30e-6±6.0e-8	130	1.08e-6±8.3e-8	4.90	1.0±0.03
F	5.4e-5±5e-5	5.31e-6±2.2e-7	155	3.62e-6±2.1e-7	5.40	1.5±0.08
g	7.2e-7±5e-8	5.69e-7±3.1e-8	70	3.09e-7±2.6e-8	1.78	1.5e-5±1e-6
h	8.4e-7±6e-8	8.57e-7±1.7e-8	74	5.01e-7±3.9e-7	2.22	1.8e-5±1e-6
i	1.0e-6±2e-7	8.87e-7±3.6e-8	73	3.16e-8±3.6e-9	2.55	8.9e-7±8e-8

5.2.1 *Intrinsic Disorder in $Na_{2-\delta}Mo_6Se_6$*

An estimate of the intrinsic disorder in each crystal is useful to compare the transport characteristics at the transition temperatures. Unfortunately, it was not possible to characterize every crystal by XRD in Grenoble for several reasons. Firstly, the number of crystals tested was too large compared to the time available at the Synchrotron beamline. Secondly, a sample for XRD analysis has to be small and thin, which is opposite to the long and large crystals required for transport measurements. Therefore, it was not possible to extract the Na vacancy value for each crystal.

However, the level of disorder can be roughly estimated from the value of $\rho(300\ K)$. XRD analysis (Sect. 4.3) indicated an Na content in the range $Na_{1.7\rightarrow 2}$. The corresponding change in carrier density would give a maximum change of ~15 % in $\rho(300\ K)$. Instead, $\rho(300\ K)$ varies by more than an order of magnitude (Table 5.1): this can only be the consequence of a disorder-induced change in the mobility. The electron mean free path must therefore vary due to the level of disorder [11].

On the other hand, the values of $\rho(T_{MI})$ and $\rho(T_{ons})$ cannot be considered as measures of disorder. At these temperatures, other electronic instabilities have to be taken into account: the situation is too complex to associate the absolute values of the resistivity purely with the level of disorder. However, it is safe to consider that all crystals are in the same metallic state at $T = 300\ K >> t_{\perp}$ with a similar level of electronic correlation [11].

5.3 Luttinger Liquid in the High Temperature Regime

The metallic region is observed at $T > t_{\perp}$. In this regime, the electrons are expected to follow a pure 1D behavior, i.e. to flow coherently along the MoSe chains. Electronic transport in 1D is described by the Luttinger liquid model, which considers

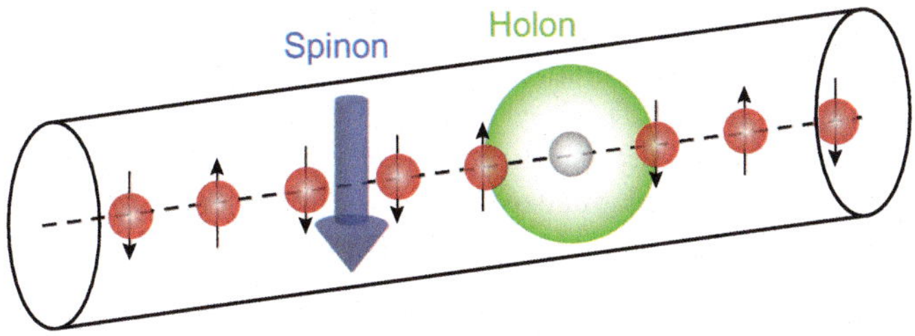

Fig. 5.3 Charge-spin separation in a Luttinger liquid. In 1D, the electron splits into two collective excitations: the holon (carrying charge) and the spinon (carrying spin). Schematic adapted from Giamarchi [15]

collective electron propagation in reduced dimension [12]. This is a major difference from the usual Fermi liquid which holds for free fermionic particles in $d > 2$ where interactions do not drastically change particle behavior.

5.3.1 Theory of the Luttinger Liquid

In 1D, electronic propagation becomes collective: an electron is obliged to interact with its neighbors in response to an applied field or current. This means that no single particle excitations (quasiparticles) exist anymore: collective excitations dominate the system and are described by Luttinger liquid (LL) theory [13, 14].

Electrons carry a spin which effectively splits from the charge due to the collective nature of the excitations, creating a charge excitation (holon, sound wave) and a spin excitation (spinon, spin wave). The holons and spinons generally have different velocities, causing the electron to separate into these two elementary excitations (Fig. 5.3).

The temperature dependance of the resistance in a clean LL is described as [16]:

$$\rho(T) \sim T^{4n\alpha} \tag{5.1}$$

where n is the filling commensurability (here $n = 1$ for the half-filled band in $Na_{2-\delta}Mo_6Se_6$), $\alpha = 2K_\rho - 3$ and K_ρ is the Luttinger parameter describing the e–e interactions. $K_\rho < 1$ corresponds to repulsive e–e interactions while attractive e–e interactions yield $K_\rho > 1$.

5.3.2 Fits to the Luttinger Liquid Model

In the presence of disorder, Eq. 5.1 is no longer valid as disorder tends to renormalize K_ρ [16]. The following aspects of the data indicate that the effect of disorder in

$Na_{2-\delta}Mo_6Se_6$ cannot be ignored [11]: the room temperature resistivity $\rho(300\ \mathrm{K})$ in Table 5.1 drastically increases from crystal A to F, and the value of T_{MI} in more disordered samples (crystals D, E, F) is higher than $t_\perp$.

Therefore, the high temperature transport data of crystals A to F presented in Fig. 5.4 are fitted using a renormalized LL model for disordered systems, Eq. 5.1, introduced in the following section.

5.3.3 Renormalized Energy Scales in a Disordered System

The LL model for clean systems (Eq. 5.1) cannot be applied to our disordered material. Instead, $\rho(T)$ in a 1D chain of spinless fermions is described by [16]:

$$\rho(T) \sim T^{\alpha} \tag{5.2}$$

where $\alpha = 2K_\rho - 2$.

LL fits to the experimental $\rho(T)$ yield $\alpha \sim 1$ in all crystals, leading to a Luttinger parameter $K_\rho \sim 3/2$ for all the crystals: this corresponds to an attractive e–e interaction [11].

A crossover from a localized to delocalized electron in the ground state has been predicted at $K_\rho = 3/2$ [16]. It was shown that for systems with $1 < K_\rho < 3/2$ the resistivity is expected to first decrease with decreasing temperature until it reaches a crossover temperature, then start increasing again. These predictions correspond very closely to our experimental data (Fig. 5.4). This suggests that electron localization might be taking place in $Na_{2-\delta}Mo_6Se_6$ at low temperature.

Dimensional crossover corresponding to an upturn in the resistivity has already been observed in other q1D materials. It has been shown in $PrBa_2Cu_4O_8$ that an unusual macroscopic localization regime develops following a dimensional crossover in the electronic state [17]. In the purple bronze $Li_{0.9}Mo_6O_{17}$, an upturn in the resistance has been widely reported [18–20] and tentatively interpreted as an electronic density wave [19]. It seems that in these q1D materials, an interesting interplay between the MI transition and the electronic dimensional crossover is taking place, although this is far from being completely understood.

The presence of e–e interactions in the LL transport renormalizes the single-particle dimensional crossover to a lower temperature $T_x < t_\perp$ [12, 21, 22]:

$$T_x \sim W \left(\frac{t_\perp}{W} \right)^{\frac{1}{1-\zeta}} \tag{5.3}$$

where W is the conduction bandwidth obtained from band structure calculations and $\zeta = \left(K_\rho + K_\rho^{-1} - 2 \right)/8 > 0$ the single particle exponent. In the case of no e–e interactions, $\zeta = 0$ and $T_x \sim t_\perp$ is obtained. As $\zeta > 0$ for a system with (attractive or repulsive) e–e interactions, T_x is always renormalized to a lower temperature.

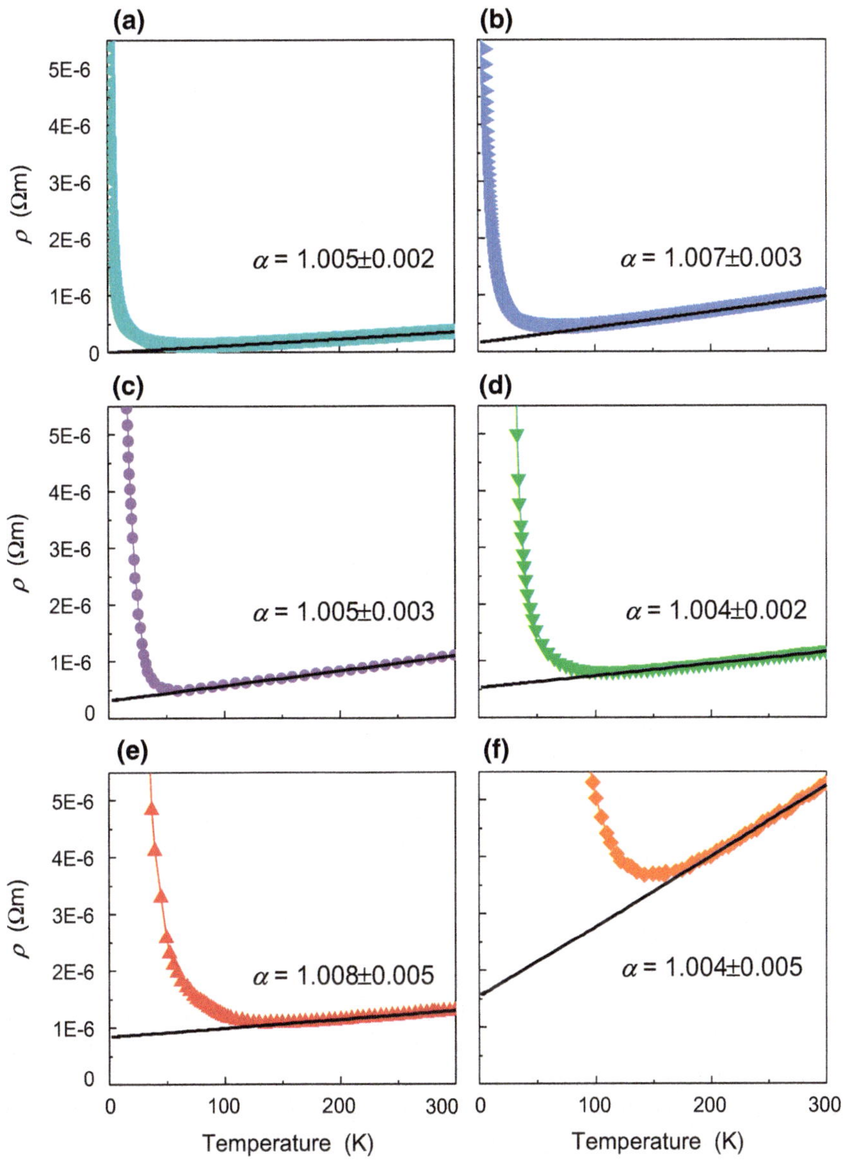

Fig. 5.4 LL fits to the temperature dependent resistivity $\rho(T)$ following Eq. 5.2 for disordered systems. The fits are performed for $\rho(T > 1.5T_{MI})$ and are shown in black line for crystals A to F in figures (**a**) to (**f**), respectively. The data sets are shown in linear graph for better visualization of the LL behavior at high temperature

This means that e–e interactions lower the temperature range over which the system exhibits 1D behavior.

The temperature of the dimensional crossover in $Na_{2-\delta}Mo_6Se_6$ using $W = 7.4$ eV (Sect. 4.4) and $K_\rho \sim 3/2$ is renormalized to $T_x \sim 102$ K [11].

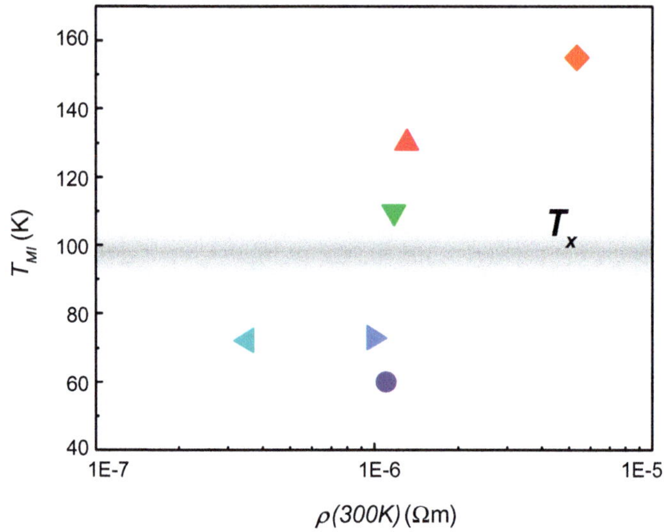

Fig. 5.5 Evolution of T_{MI} with ρ(300 K). The horizontal grey shading indicates the renormalized dimensional crossover temperature $T_x \sim 102$ K

The values of T_{MI} are plotted as a function of ρ(300 K) in Fig. 5.5. Interestingly, the position of T_x indicates a clear separation between crystals A, B, C and crystals D, E, F.

Crystals A, B, C have a relatively similar $T_{MI} < T_x$. On the other hand in crystals D, E, F, the value of T_{MI} rises and lies at $T_{MI} > T_x$. In these last crystals, the high value of T_{MI} cannot be explained by e–e interaction renormalization. Instead, the presence of a large level of disorder must be taken into account.

In a highly disordered LL featuring an attractive e–e interaction ($K_\rho > 1$), a minimum has been predicted to develop in $\rho(T)$ at a charge pinning temperature T_{pin} [16]. This temperature represents a threshold where transport in a LL close to half-filling (such as $Na_{2-\delta}Mo_6Se_6$) behaves similarly to that in an electronic CDW pinned by disorder. Consequently, $\rho(T)$ begins to diverge despite the absence of a phase transition. In the highly disordered crystals, a dimensional crossover is still expected to take place at T_x. However, the resistivity starts to diverge at higher temperature due to the larger levels of disorder in the system, masking the dimensional crossover in the transport measurements [11].

The transport behavior in the high temperature regime is therefore quantitatively compatible with a disordered Luttinger liquid with attractive e–e interactions as indicated by $K_\rho > 1$ [11].

It is important to note that at the time of this work, the theory of disordered LL is largely incomplete, and the theory of disordered LL with attractive e–e interactions is hardly explored [12, 16].

5.4 Metal-Insulator Transition in the Intermediate Temperature Regime

As seen in Fig. 5.1, a divergence of the resistance dominates the intermediate temperature region (~150−6 K, regime 2 in Fig. 5.2). Upturns or divergence in $\rho(T)$ have been observed in q1D materials and attributed to electron localization [17, 23–25], LL physics [26], density wave formation [19, 27, 28], electron density fluctuations [29], or proximity to Mott instabilities [30].

The information available for $Na_{2-\delta}Mo_6Se_6$ and the behavior of the transport data are suggestive of the following: first, below $T_x \sim 102$ K the electronic transport is 3D (although still anisotropic), making LL physics an unlikely possibility; second, $\rho(T)$ in $Na_{2-\delta}Mo_6Se_6$ presents a broad minimum which contrasts strongly with the abrupt jumps in $\rho(T)$ for nesting-driven density waves as reported in q1D, for example $NbSe_3$ [27].

5.4.1 Mechanisms for the Metal-Insulator Transition

Thermal Activation

A phase transition opening an energy gap E_g in the DoS at the Fermi energy exhibits thermally-activated transport following the Arrhenius equation:

$$\rho(T) \propto \exp\left[\frac{E_g}{k_B T}\right] \tag{5.4}$$

Mott insulators and fully gapped density wave systems exhibit Arrhenius behavior in their electrical transport.

In $Na_{2-\delta}Mo_6Se_6$, it is safe to consider that any gap opening would lead to a fully gapped Fermi surface: the Fermi sheets are planar and highly-nested (Fig. 4.6) [11].

Weak Localization

In the low disorder limit, quantum interference between impurity-scattered wavefunctions enhances back-scattering even for delocalized electrons: this is known as weak localization. The electron transport is controlled by inelastic scattering events, which leads a logarithmically-divergent resistivity:

$$\rho(T) = \rho_0 + \frac{\rho_1}{ln(T)} \tag{5.5}$$

where ρ_0 and ρ_1 are constants.

Weak localization is a precursor to Anderson localization (strong localization), and takes place at low levels of disorder.

Anderson Localization

In a strongly disordered system, Anderson localization takes place (Sect. 2.4.1). Transport of localized electrons is generally enabled by phonon-mediated hopping between nearby localized states [31]. This is described by the variable range hopping (VRH) model:

$$\rho(T) = \rho_0 \exp\left[\left(\frac{T_0}{T}\right)^{\frac{1}{1+d}}\right] \tag{5.6}$$

where T_0 is the characteristic localization temperature and d is the dimensionality. d reflects the effective dimensionality of the system and does not accurately correspond to the electronic dimensionality. In $Na_{2-\delta}Mo_6Se_6$ crystals for example, the electronic transport is anisotropic 3D below $T_x \sim 102$ K. However, due to the large structural anisotropy and the electron hopping energy ratio, the effective dimensionality is expected to resemble that of a q1D system, i.e. $1 < d < 2$ [11].

In 1D materials fitted with VRH, $d = 1$ also corresponds to Efros–Shklovskii VRH for a system of arbitrary dimensionality, describing the opening of a soft gap at the Fermi level due to Coulomb repulsion [32].

The MI transitions of the least (A) and most (F) disordered crystals are fitted to each model in Fig. 5.6. All proportionality constants, exponents, and scaling parameters (T_0, d, E_g, etc.) are kept as completely free variables during the least-squares fitting procedure.

The fits to thermal activation and weak localization fail to describe the upturn in the resistivity. In contrast, the fits to the VRH model nicely follow the data. A standard Pearson χ^2 test is used to compare the quality of each fit. For both crystals, the fits to VRH presents a χ^2 value at least two orders of magnitude lower than the other models.

This fitting exercise suggests that as the temperature drops, electrons in $Na_{2-\delta}Mo_6Se_6$ are strongly localized although no gap develops at the Fermi energy [11].

A more disordered system exhibits states which are more localized in real space, leading to an increased resistance. The extracted T_0 from the VRH model is proportional to the separation between E_F and the mobility edge. Therefore, larger level of disorder yields larger T_0.

Each $Na_{2-\delta}Mo_6Se_6$ crystal is now analyzed to compare the level of disorder using the VRH fitting procedure.

5.4.2 *Fits to the Variable Range Hopping Model*

Fits to the VRH model for $\rho(T < T_x)$ are presented in Figs. 5.7 and 5.8 for all the measured crystals: A to F and g to i, respectively. A zoom onto the lowest temperature behavior of crystals g and h is included, highlighting the presence of a peak at T_{ons}. The quality of the fits is confirmed in the insets showing $\rho(T^{1/(1+d)})$ on a semi-logarithmic scale.

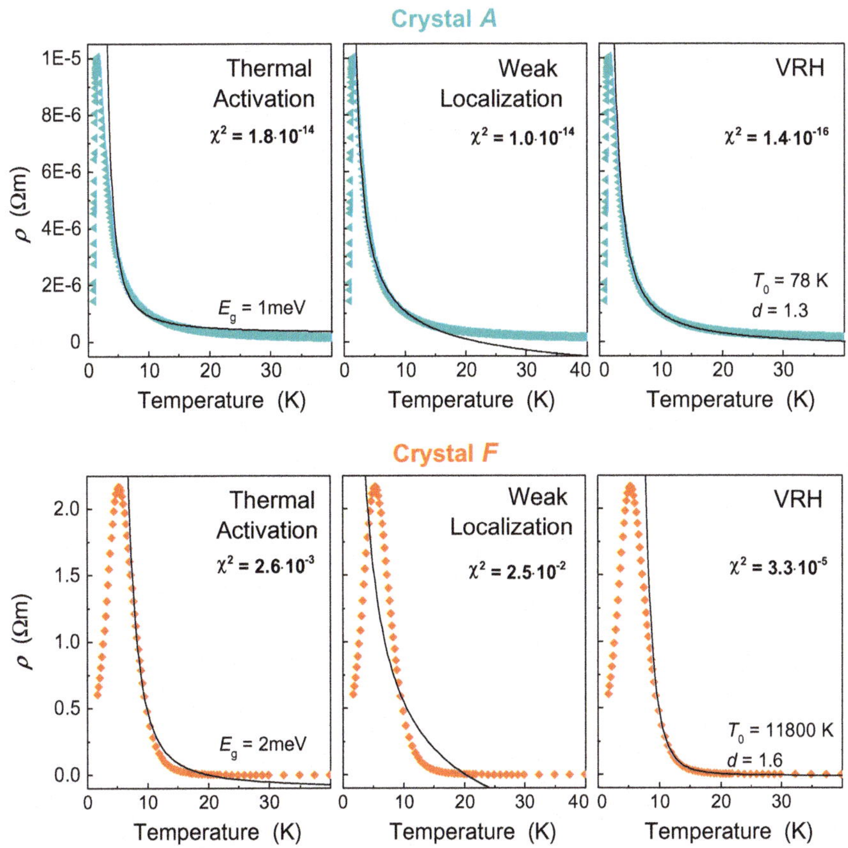

Fig. 5.6 Fits of MI transitions to the thermal activation, weak localization, and Anderson localization models for the least (A) and most (F) disordered crystals

Fits to $\rho(T)$ and linear trends in $\rho(T^{1/(1+d)})$ in all crystals indicate that the insulating behavior observed at the intermediate temperature regime is controlled by VRH, due to Anderson localization [11].

The fitting parameters T_0 and d extracted from the VRH model are presented in Table 5.2. The electronic transport in $Na_{2-\delta}Mo_6Se_6$ is 3D at $T < T_x$, where T_x replaces $t_\perp$ in the model of a q1D system. However, macroscopically, the crystalline structure of the system remains q1D. The values of d for all crystals reflects the structural dimension: d lies between 1.2 and 1.7 (Table 5.2). In arrays of disordered conducting chains, d was predicted to be 1.5 [33]. The experimental data are therefore in good agreement with theoretical predictions. As $d > 1$ is always observed, any Coulomb gap opening due to Efros–Shklovskii VRH can be discarded.

Earlier in this chapter, the values of ρ(300K) were presented as a good measure of disorder. The relationship between T_0 and ρ(300K) is shown in Fig. 5.9: from A to F (including g, h and i), T_0 and ρ(300K) monotonically increase. This supports that ρ(300K) and T_0 both represent the level of disorder [11].

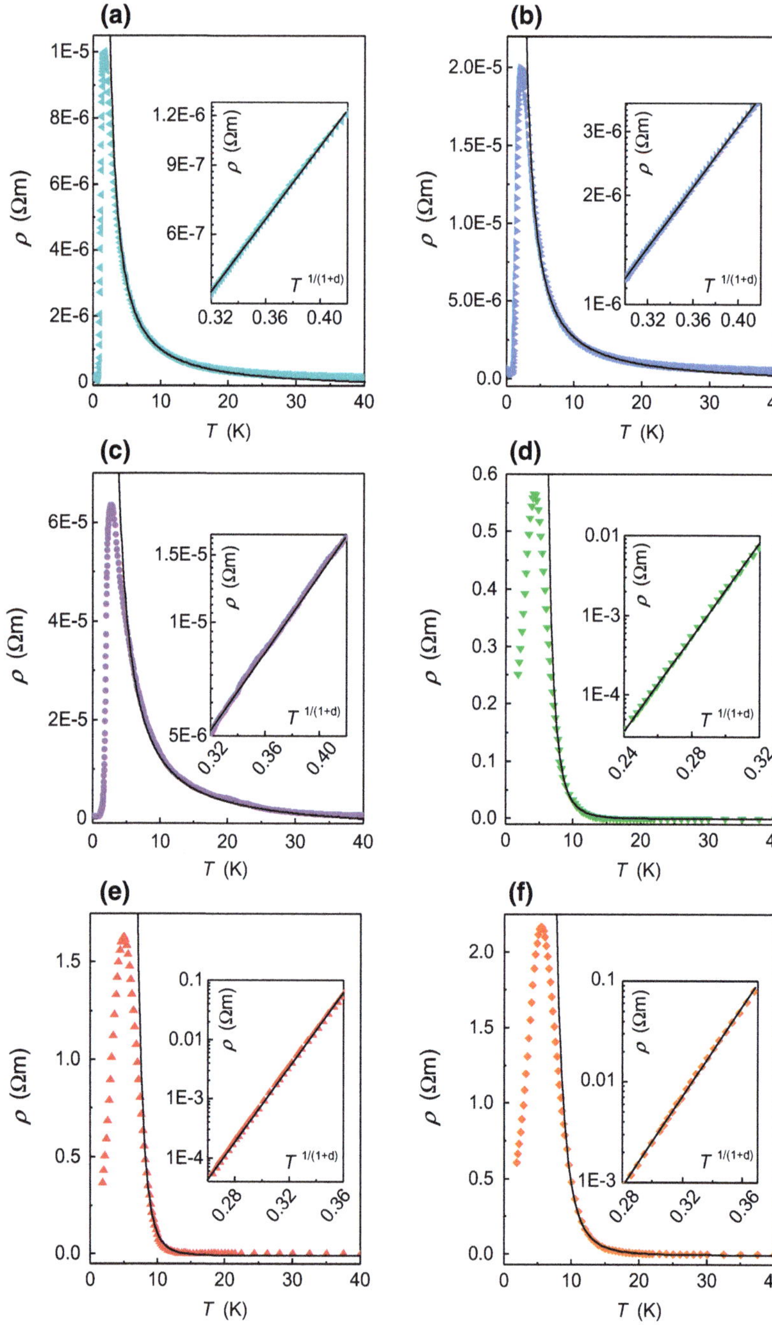

Fig. 5.7 Intermediate temperature transport exhibiting Anderson localization in crystals A to F: fits to Eq. 5.6 for $\rho(T < T_x)$ are shown as black lines. **Insets**: $\rho(T^{1/(1+d)})$ plotted on a semi-logarithmic scale highlights the goodness of the fits

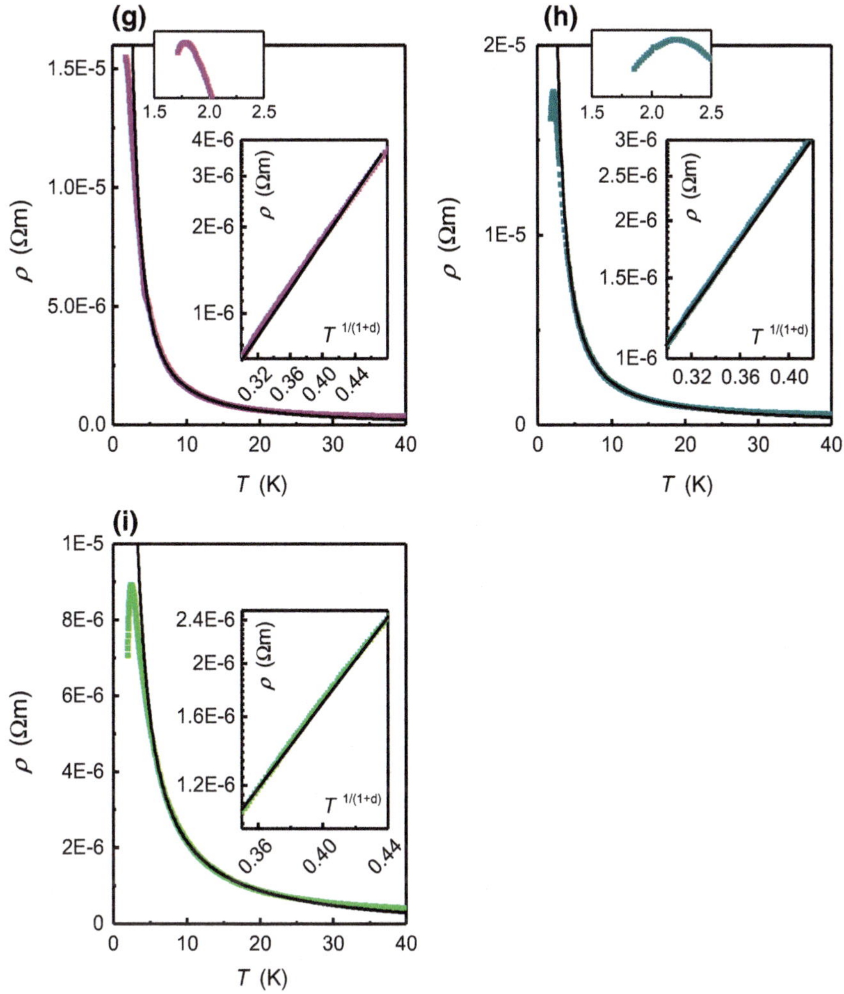

Fig. 5.8 Intermediate temperature transport exhibiting Anderson localization in crystals g to i: fits to Eq. 5.6 for $\rho(T < T_x)$ are shown as black lines. For crystals g and h a zoom highlights the peaks at T_{ons}. **Insets**: $\rho(T^{1/(1+d)})$ plotted on a semi-logarithmic scale highlights the goodness of the fits

5.4.3 Signatures of Anderson Localization

The VRH fits indicate the role of Anderson localization in the intermediate temperature regime and confirm the increase of disorder from crystal A to F [11]. In addition, signatures of strong electron localization can be tracked in the frequency dependence of the ac conductivity $\sigma(\omega)$, magnetoresistance $R(H)$, and magnetization measurements.

Table 5.2 VRH fitting range and parameters for all measured crystals

Crystal	Fitting range [K]	T_0 [K]	d [–]
A	4.5–18	78±12	1.3±0.1
B	4.5–18	112±21	1.4±0.1
C	6–28	180±30	1.5±0.1
D	7.5–27	2670±990	1.2±0.1
E	8–20	6500±1050	1.3±0.1
F	9–27	11800±3500	1.6±0.1
g	4.5–18	108±22	1.3±0.1
h	4.5–18	129±31	1.4±0.1
i	5–20	134±36	1.7±0.1

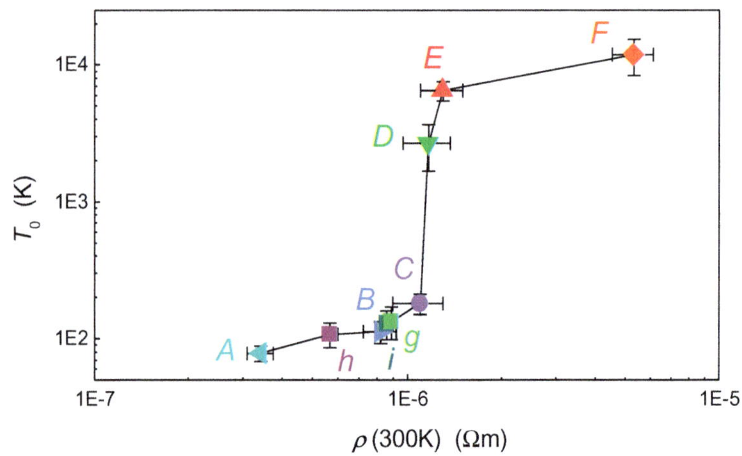

Fig. 5.9 Disorder in $Na_{2-\delta}Mo_6Se_6$: Characteristic localization temperature T_0 versus ρ(300 K) for all measured crystals

Frequency Dependence of the Ac Conductivity $\sigma(\omega)$

In a strongly-localized material, $\sigma(\omega)$ is expected to display a dependence on ω even for small frequencies. At zero temperature and in the limit $\omega \to 0$ this relationship derived by Mott follows [31]:

$$\text{Re}\{\sigma(\omega)\} \sim N(E_F)^2 l^{d+2} \omega^2 \left| \log\left(\frac{1}{\omega}\right)^{\alpha} \right| \tag{5.7}$$

where $N(E_F)$ is the DoS at the Fermi level, l is the localization length, d the dimensionality and α an exponent related to the dimensionality. It was originally shown that $\alpha = d + 1$; however recent work considering the Anderson transition suggests

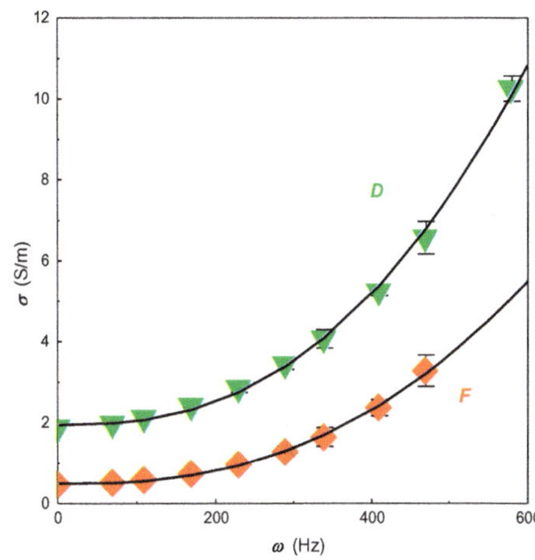

Fig. 5.10 Frequency dependence of the ac conductivity $\sigma(\omega)$ in crystal D and F. The black lines are the fits to Eq. 5.7 using $\alpha = 3$. The data were acquired at $T = 4.6$ and 6 K, respectively

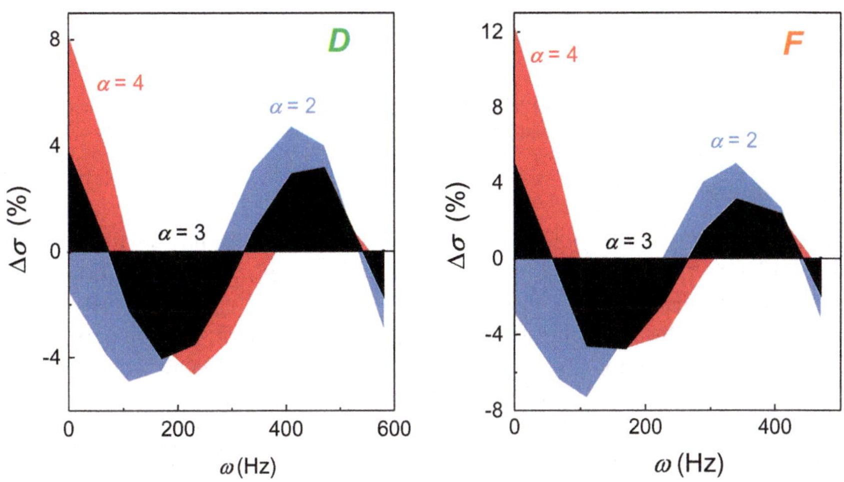

Fig. 5.11 Percentage errors $\Delta\sigma = \left(\sigma_{fit} - \sigma_{experiment}\right)/\sigma_{experiment}$ in the fits to $\sigma(\omega)$ as a function of the frequency ω for $\alpha = 1, 2, 3$

that $\alpha = d + 2$ [34]. Figure 5.10 presents the fits to the highest disordered crystals[1] and confirms that Eq. 5.7 describes the experimental data well [11]. The data were acquired at the lowest temperature achievable while staying in the localized regime, i.e. at $T > T_{ons}$.

The fitting procedure was performed for different α values. Figure 5.11 indicates the percentage errors in the fits as a function of frequency for $\alpha = 1, 2, 3$. This exercise indicates a superior fit for $\alpha = d + 2 = 3$.

[1]Crystal E broke due to repeated thermal cycles before a full $\sigma(\omega)$ dataset could be completed.

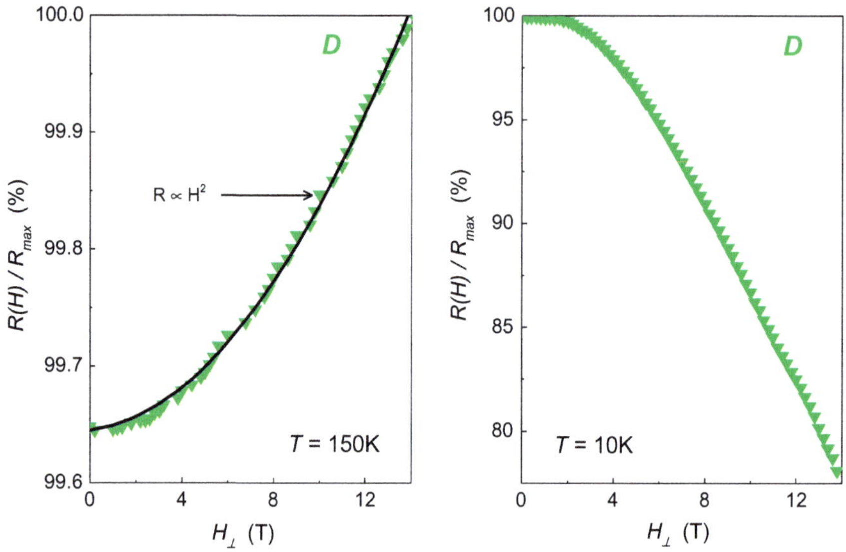

Fig. 5.12 Normalized magnetoresistance $R(H)/R_{max}$ as a function of perpendicular magnetic field $H_{\perp}$ in crystal D at $T = 150$ and 10 K

Magnetoresistance $R(H)$

The magnetoresistance in the intermediate temperature regime is presented in Fig. 5.12 for crystal D. At $T = 150$ K, (in the metallic phase $dR/dT > 0$) the effects of disorder are weak and $R \propto H^2$ as expected for an open Fermi surface. At $T = 10$ K (in the insulating phase $dR/dT < 0$) the data show a drop of over 20% from zero-field value for a perpendicular field of 13 T. This behavior corresponds to a delocalization of electronic states and has been predicted for systems in the Anderson localization regime [35]. An applied magnetic field induces Zeeman splitting, hence reduces the localization energy for half the electrons leading to a fall in the measured resistance.

If a Coulomb gap were present at E_F, no such delocalization would be possible and the magnetoresistance would be positive as per the usual Efros-Shlovskii VRH order [32]. The large variation in $R(H)$ between 150 and 10 K correlates with a crossover from metallic behavior to strong localization [11].

Magnetization and Susceptibility χ

In the high temperature regime, $M_{2-\delta}Mo_6Se_6$ are weakly diamagnetic and their dimensionless susceptibility χ is temperature-invariant [36, 37]. The magnetization of $Na_{2-\delta}Mo_6Se_6$ was measured from room temperature to 1.8 K, the lowest achievable temperature for our SQUID magnetometer. The left graph in Fig. 5.13 shows a weak paramagnetism emerging below T_x. This is due to strong localization within this temperature regime. When the temperature is decreased and crosses $T_x \sim 102$ K, electrons start to localize: weak diamagnetism vanishes and Curie paramagnetism emerges from localized electrons showing a $1/T$ dependance [11]. Such behavior was previously reported in other Anderson localized materials [38].

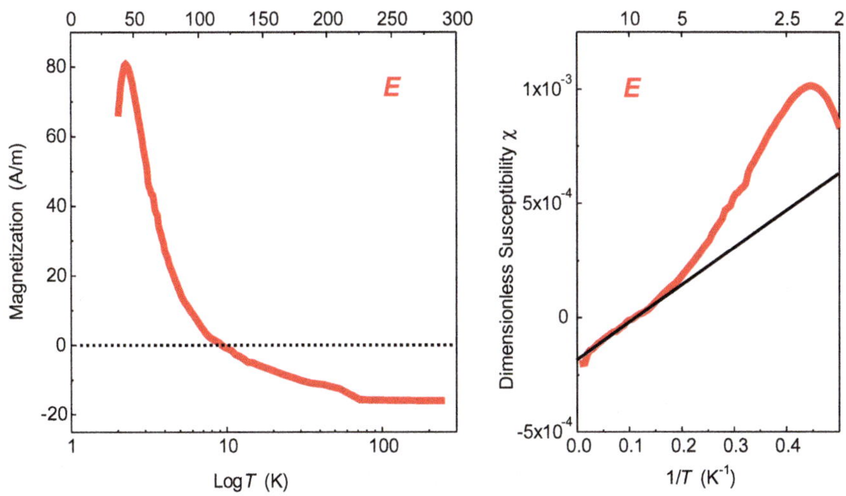

Fig. 5.13 Signatures of electron localization in the magnetization in crystal E. **Left**: Magnetization as a function of T on a semi-logarithmic scale, extracted after careful background subtraction. The dashed line indicates the zero value. **Right**: Dimensionless susceptibility χ as a function of $1/T$. The black straight line is just a guide to the eye. Data acquired by A.P. Petrović

Furthermore, χ is presented in the right panel of Fig. 5.13. The behavior of χ as a function of $1/T$ is non-linear with $d^2\chi/d(1/T)^2 > 0$ as highlighted with the black straight line. This implies that the paramagnetic spin density increases with decreasing temperature. If paramagnetism were caused by paramagnetic impurities, the spin density would be expected to follow a linear $\chi \propto 1/T$ behavior. Furthermore, χ is constant at higher temperature, although paramagnetism only emerges at $T < T_x$: paramagnetic impurities would be visible at all temperatures. Finally, any SDW at low temperature would create a peak in χ due to anti-ferromagnetic ordering. The only visible peak corresponds to T_{ons} and is due to emergent superconductivity (which will be studied in the next chapter).

5.4.4 Emergence of a Mobility Edge

It has been shown that all crystals exhibit strong localization and VRH transport in the divergent resistivity region. However, $\rho(T_{ons})$ increases by 4 orders of magnitude between crystal C and D (Fig. 5.14). This is observed by a step-like feature in Figs. 5.9 and 5.14. Such behavior is typically associated with the crossing of the mobility edge, or critical disorder, in 3D materials. The data are therefore suggestive of a crossover to strong localization and the presence of a "q1D mobility edge" [11].

In 3D materials, the theoretical understanding of the mobility edge is well developed [39]. However, experimental observations of its fundamental nature have only

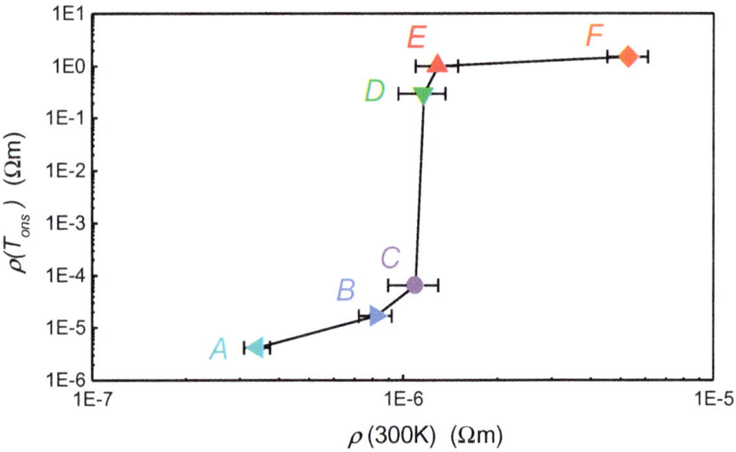

Fig. 5.14 Evolution of $\rho(T_{ons})$ as a function of $\rho(300\text{ K})$ for crystals A to F

been reported recently [40]. A mobility edge theoretically does not exist in 1D as all states are localized for any level of disorder. Nonetheless, the question of whether a mobility edge can develop in q1D materials remains open: once the system has crossed its dimensional crossover temperature T_x, the electrons are in an anisotropic 3D state. The observation of a mobility edge in $Na_{2-\delta}Mo_6Se_6$ suggests that disordered q1D systems can indeed exhibit a transition from extended to localized states with increasing disorder, similar to 3D systems.

Other signatures in the transport data are supportive of a mobility edge between crystals C and D. First, the Mott frequency dependence $\sigma(\omega) \propto \omega^2 log^{\alpha}(1/\omega)$ (Eq. 5.7) is not observed in the least disordered crystals (Fig. 5.15). This confirms that only crystals D to F lie on the insulating side of the mobility edge and exhibit completely localized electronic wavefunction.

Secondly, the evolution of T_{MI} (Fig. 5.5) from $T_{MI} \approx T_x$ in crystals A, B, C to the disorder-dependent $T_{MI} = T_{pin}$ in crystals D, E, F indicates that a critical disorder level is driving the emergence of T_{pin}.

Finally, the value of the extracted $K_\rho \sim 3/2$ for all crystals using the disordered LL model corresponds closely to the predicted crossover from localized to delocalized electrons [16]. This suggests a potential relationship between the dimensional crossover and the presence of a mobility edge [11].

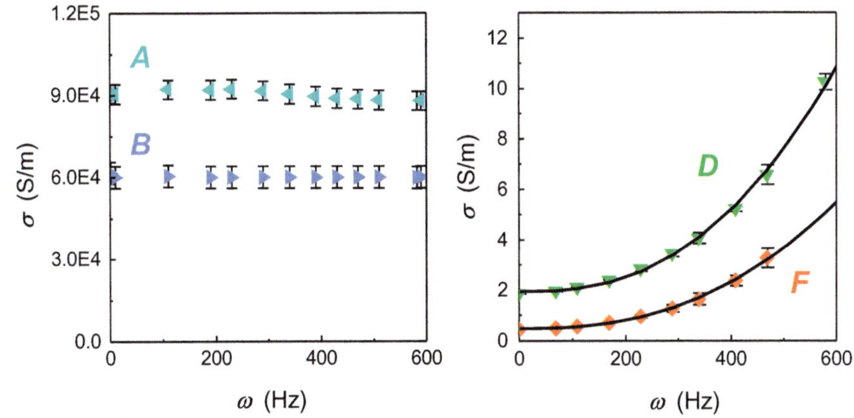

Fig. 5.15 Contrast in $\sigma(\omega)$ between the least and most disordered crystals. The data was taken at $T = 4.9, 4.9, 4.6, 6$ K for crystals A, B, D and F, respectively. The graph in the right hand panel is similar to Fig. 5.10

5.5 Concluding Remarks

The electronic normal state in $Na_{2-\delta}Mo_6Se_6$ is now better understood. At high temperature, it is a disordered LL with attractive e–e interactions. With decreasing temperature, a dimensional crossover from 1D to anisotropic 3D takes place and $t_\perp \approx 120$ K $\longrightarrow T_x \approx 102$ K due to disorder.

A further decrease in temperature causes a divergence in the resistivity due to electron localization. The transport data are described by the VRH model, while signatures of strong localization are present in magnetoresistance data, the Mott frequency dependence of the ac conductivity, and magnetization.

Moreover, a mobility edge is observed in several different measurements between less and more disordered crystals. This suggests that even though $Na_{2-\delta}Mo_6Se_6$ is q1D, critical disorder induces a crossover in the electron wavefunctions from an extended to localized state. Such behavior in low-dimensional systems is largely unexplored.

References

1. V.F. Gantmakher, V.T. Dolgopolov, Superconductor-insulator quantum phase transition. Physics-Uspekhi **53**(1), 1–49 (2010)
2. M. Ma, P.A. Lee, Localized superconductors. Phys. Rev. B **32**(9), 5658–5667 (1985)
3. M.V. Sadovskii, Localization effects in high-temperature superconductors. Ann. New York Acad. Sci. **581**(213), 207–216 (1990)
4. L.N. Bulaevskii, M.V. Sadovskii, Anderson localization and superconductivity. J. Low Temp. Phys. **59**(1/2), 89–113 (1985)

5. L.N. Bulaevskii, S.V. Panyukov, M.V. Sadovskii, Inhomogeneous superconductivity in disordered metals. Zh. Eksp. Teor. Fiz. **92**(2), 672–688 (1987)
6. A.M. Finkel'stein, Suppression of superconductivity in homogeneously disordered systems. Phy. B Conden. Matter **197**(1–4), 636–648 (1994)
7. Y. Dubi, Y. Meir, Y. Avishai, Nature of the superconductor-insulator transition in disordered superconductors. Nature **449**(7164), 876–80 (2007)
8. B. Sacépé, C. Chapelier, T. Baturina, V. Vinokur, M. Baklanov, M. Sanquer, Disorder-induced inhomogeneities of the superconducting state close to the superconductor-insulator transition. Phys. Rev. Lett. **101**(15), 157006 (2008)
9. B. Sacépé, T. Dubouchet, C. Chapelier, M. Sanquer, M. Ovadia, D. Shahar, M. Feigel'man, L. Ioffe, Localization of preformed Cooper pairs in disordered superconductors. Nat. Phys. **7**(3), 239–244 (2011)
10. C. Brun, T. Cren, V. Cherkez, F. Debontridder, S. Pons, D. Fokin, M.C. Tringides, S. Bozhko, L.B. Ioffe, B.L. Altshuler, D. Roditchev, Remarkable effects of disorder on superconductivity of single atomic layers of lead on silicon. Nat. Phys. **10**(June), 444–450 (2014)
11. A.P. Petrović, D. Ansermet, D. Chernyshov, M. Hoesch, D. Salloum, P. Gougeon, M. Potel, L. Boeri, C. Panagopoulos, A disorder-enhanced quasi-one-dimensional superconductor. Nat. Commun. **7**, 12262 (2016)
12. T. Giamarchi, Theoretical framework for quasi-one dimensional systems. Chem. Rev. **104**(11), 5037–5056 (2004)
13. S.-I. Tomonaga, Remarks on Bloch's Method of Sound Waves Applied to Many-fermion Problems (1950). ISSN 0033-068X
14. J.M. Luttinger, An exactly soluble model of a many-fermion system. J. Math. Phys. **4**(9), 1154 (1963)
15. T. Giamarchi, Deconstructing the electron. Physics **2**(78) (2009)
16. T. Giamarchi, *Quantum Physics in One Dimension* (Clarendon Press, Oxford, 2003)
17. A. Narduzzo, A. Enayati-Rad, S. Horii, N.E. Hussey, Possible coexistence of local itinerancy and global localization in a quasi-one-dimensional conductor. Phys. Rev. Lett. **98**(14), 1–4 (2007)
18. Y. Matsuda, M. Sato, M. Onoda, K. Nakao, On the anomalous transport properties of Li0.9Mo6O17. J. Phys. C Solid State Phys. **19**(30), 6039–6052 (1986)
19. X. Xu, A.F. Bangura, J.G. Analytis, J.D. Fletcher, M.M.J. French, N. Shannon, J. He, S. Zhang, D. Mandrus, R. Jin, N.E. Hussey, Directional field-induced metallization of quasi-one-dimensional Li0.9Mo6O17. Phys. Rev. Lett. **102**(20), 1–4 (2009)
20. J.-F. Mercure, A.F. Bangura, X. Xu, N. Wakeham, A. Carrington, P. Walmsley, M. Greenblatt, N.E. Hussey, Upper critical magnetic field far above the paramagnetic pair-breaking limit of superconducting one-dimensional Li0.9Mo6O17 single crystals. Phys. Rev. Lett. **108**(18), 187003 (2012)
21. D. Boies, C. Bourbonnais, A.-M.S. Tremblay, One-particle and two-particle instability of coupled Luttinger liquids. Phys. Rev. Lett. **74**(6), 968–971 (1995)
22. C. Bourbonnais, L.G. Caron, The role of kinetic interchain coupling in quasi-1D conductors. Physica B+C **143**(1–3), 450–452 (1986)
23. M. Sato, Y. Matsuda, H. Fukuyama, Localisation and superconductivity in Li0.9Mo6O17. J. Phys. C: Solid State Phys. **20**, L137–L142 (1987)
24. A. Enayati-Rad, A. Narduzzo, F. Rullier-Albenque, S. Horii, N.E. Hussey, Irradiation-induced confinement in a quasi-one-dimensional metal. Phys. Rev. Lett. **99**(13), 1–4 (2007)
25. S. Khim, B. Lee, Ki.Y. Choi, B.-G. Jeon, D.H. Jang, D. Patil, S. Patil, R. Kim, E.S. Choi, S. Lee, J. Yu, K.H. Kim, Enhanced upper critical fields in a new quasi-one-dimensional superconductor Nb2PdxSe5. New J. Phys. **15**(12), 123031 (2013)
26. C.A.M. dos Santos, M.S. Da Luz, Y.-K. Yu, J.J. Neumeier, J. Moreno, B.D. White, Electrical transport in single-crystalline Li0.9Mo6O17: a two-band Luttinger liquid exhibiting Bose metal behavior. Phys. Rev. B - Conden. Matter Mater. Phys. **77**(19), 1–4 (2008)
27. N.P. Ong, P. Monceau, Anomalous transport properties of a linear-chain metal: NbSe3. Phys. Rev. B **16**(8), 3443–3455 (1977)

28. C.A.M. dos Santos, B. White, Yi.-K. Yu, J. Neumeier, J. Souza, Dimensional crossover in the purple bronze Li0.9Mo6O17. Phys. Rev. Lett. **98**(26), 266405 (2007)
29. A.P. Petrović, R. Lortz, G. Santi, M. Decroux, Phonon mode spectroscopy, electron-phonon coupling, and the metal-insulator transition in quasi-one-dimensional M2Mo6Se6. Phys. Rev. B **82**(23), 235128 (2010)
30. P. Chudzinski, T. Jarlborg, T. Giamarchi, Luttinger-liquid theory of purple bronze Li0.9Mo6O17 in the charge regime. Phys. Rev. B **86**(7): 075147 (2012)
31. N.F. Mott, Conduction in non-crystalline materials: III. Localized states in a pseudogap and near extremities of conduction and valence bands. Philos. Mag. **19**(160), 835–852 (1969)
32. A.L. Efros, B.I. Shklovskii, Coulomb gap and low temperature conductivity of disordered systems. J. Phys. C: Solid State Phys. **8**(4), L49 (1975)
33. M.M. Fogler, S. Teber, B.I. Shklovskii, Variable-range hopping in quasi-one-dimensional electron crystals. Phys. Rev. B **69**(3), 035413 (2004)
34. A. Klein, O. Lenoble, P. Müller, On Motts formula for the ac-conductivity in the Anderson model. Ann. Math. **166**(2), 549–577 (2007)
35. H. Fukuyama, K. Yoshida, Negative magnetoresistance in the Anderson localized states. J. Phys. Soc. Jpn **46**(1), 102–105 (1979)
36. J.M. Tarascon, G.W. Hull, F.J. DiSalvo, A facile synthesis of pseudo one-monodimensional ternary molybdenum chalcogenides M2Mo6X6. Mater. Res. Bull. **19**(7), 915–924 (1984a)
37. A.P. Petrović, R. Lortz, G. Santi, C. Berthod, C. Dubois, M. Decroux, A. Demuer, A.B. Antunes, A. Paré, D. Salloum, P. Gougeon, M. Potel, Ø. Fischer, Multiband superconductivity in the Chevrel phases SnMo6S8 and PbMo6S8. Phys. Rev. Lett. **106**(1), 017003 (2011)
38. R. Inada, O. Yoshichika, T. Sei-ichi, Paramagnetic property in the commensurate charge density wave phase of 1T-TaS2. J. Phys. Soc. Jpn **50**(4), 1217–1221 (1981)
39. F. Evers, A. Mirlin, Anderson transitions. Rev. Modern Phys. **80**(4), 1355–1417 (2008)
40. G. Semeghini, M. Landini, P. Castilho, S. Roy, G. Spagnolli, A. Trenkwalder, M. Fattori, M. Inguscio, G. Modugno, Measurement of the mobility edge for 3D Anderson localization. Nat. Phys. **11**(7), 554–559 (2015)

Chapter 6
Superconducting Transition and Pairing Enhancement by Disorder

6.1 Introduction

The presence of a superconducting ground state in $Tl_{2-\delta}Mo_6Se_6$ and $In_{2-\delta}Mo_6Se_6$ [1–3] suggests that in $Na_{2-\delta}Mo_6Se_6$, the peak in $\rho(T)$ at T_{ons} may be due to the onset of superconductivity. In Chap. 5, the normal state transport was shown to be dominated by VRH transport due to strong electron localization. Nonetheless, the finite resistance measured at $T \geq T_{ons}$ in each crystal indicates that a superconducting instability can develop within the system.

In filamentary materials, phase-coherent superconductivity develops between 1D filaments via Josephson coupling, occurring at temperature T_J. The value of T_J can be estimated from the electron hopping energies: $T_J \geq t_\perp^2/t_\parallel$ (Sect. 2.3). For most q1D superconductors, the critical pairing temperature T_c is much lower than $t_\perp^2/t_\parallel$ and the superconducting transition takes place from an anisotropic 3D normal state. Nevertheless, if the anisotropy of the system and the pairing interaction are both sufficiently large, the onset temperature for superconducting fluctuations T_{ons} may be greater than $t_\perp^2/t_\parallel$: in this case, a 2-step superconducting transition occurs. Below T_{ons}, 1D superconductivity develops within individual chains and is dominated by phase slips (Sect. 2.2). At lower temperature, transverse Josephson coupling establishes phase coherence at $T_J < T_{ons}$: a Meissner state develops and the electrical resistance falls to zero. In $Na_{2-\delta}Mo_6Se_6$, the second scenario applies as $t_\perp^2/t_\parallel \sim 1$ K and $T_{ons} = 1.68 \rightarrow 5.4$ K (Tables 4.1, 5.1) [4, 5].

Indications of strongly correlated electron physics in the normal state (attractive Luttinger liquid, Anderson localization) suggest that an emergent superconducting state is likely to be atypical. Trends in the transport data between the different crystals have shown that interesting low temperature phenomena are involved: first, the shapes of the superconducting transition vary strongly between the least disordered (A, B, C) and most disordered (D, E, F) crystals; second, the value of T_{ons} evolves strongly between crystals.

The present chapter discusses the experimental study and the associated analysis: globally-coherent superconductivity is shown to emerge from the coupling of 1D

D. Ansermet, *Emergent Superconductivity in Low Dimensions*, Springer Theses,
https://doi.org/10.1007/978-981-13-2941-8_6

fluctuating filaments. Additionally, the presence of disorder is shown to give rise to an enhancement in T_{ons}, T_J and the critical magnetic field H_{c2}. A theoretical justification is proposed to explain these observations. This work was published in 2016 in *ACS nano* [4] and *Nature Communications* [5].

6.2 Identifying Superconductivity in $Na_{2-\delta}Mo_6Se_6$

We know from Chap. 2 that a superconducting state exhibiting global phase coherence depicts two main signatures in the electronic transport: zero resistance and an emergent Meissner state. To investigate whether the observed peak at T_{ons} is of superconducting origin, these two signatures are tracked in the measured crystals.

The low temperature behavior of $\rho(T)$ is depicted in Fig. 6.1 (corresponding to Fig. 5.1 for a smaller T range). In crystals A, B, C the resistance is measured down to 50 mK using the dilution refrigerator: the drop below T_{ons} is sharp and $\rho(T \to 0) \sim 0$. In a filamentary material such as $Na_{2-\delta}Mo_6Se_6$, some parts of the sample may remain in the normal state due to inhomogeneity leading to a finite resistivity at $T = 0$. In the data, the resistivity drops by at least 4 orders of magnitude below T_{ons} and stabilizes close to zero at the lowest temperature. This indicates a phase-coherent superconducting ground state [4, 5].

In contrast, for crystals D, E, F, transport measurements were performed down to ~ 1.8 K in the PPMS and show $\rho(T \to 0) \neq 0$: fluctuations in the superconducting state and inhomogeneity are important in these more disordered crystals.

The emergence of a Meissner state is examined using the SQUID magnetometer with base temperature $T = 1.7$ K. Due to the attainable temperature range, only crystals D and E have been measured.[1] Figure 6.2 presents the results of this investigation: a small Meissner state is detected in the zero-field-cooled/field-cooled (ZFC/FC) thermal cycle of the dimensionless susceptibility χ. Comparing the position of T_{ons} and the emergence of the Meissner state indicates that electron pairing and local 1D coherence develop at higher temperature than any transverse phase coherence: $T_{ons} > T_{\text{Meissner effect}}$. The observation of a Meissner state indicates that transverse phase coherence is established at least within certain regions of the crystals [5].

An ideal Meissner state would yield $\chi = -1$ (perfect diamagnetism), while the data indicate $\chi = -10^{-3} \to -10^{-4}$. This corresponds to superconducting volume fraction of $< 0.1\%$ for these crystals [5]. In q1D materials, the large penetration depth λ normal to the 1D axis yield measured superconducting volumes $<100\%$ [2]. However, the significantly lower value measured here is due to an effect of the large inhomogeneity (disorder) in the crystals: emergent spatial inhomogeneity is intrinsic to strongly-localized superconductors [6–9].

This suggests that in highly disordered crystals such as D and E, coherent superconductivity is only established within small regions of the crystals [5].

[1] Crystal F could not be measured due to a fracture in the crystal.

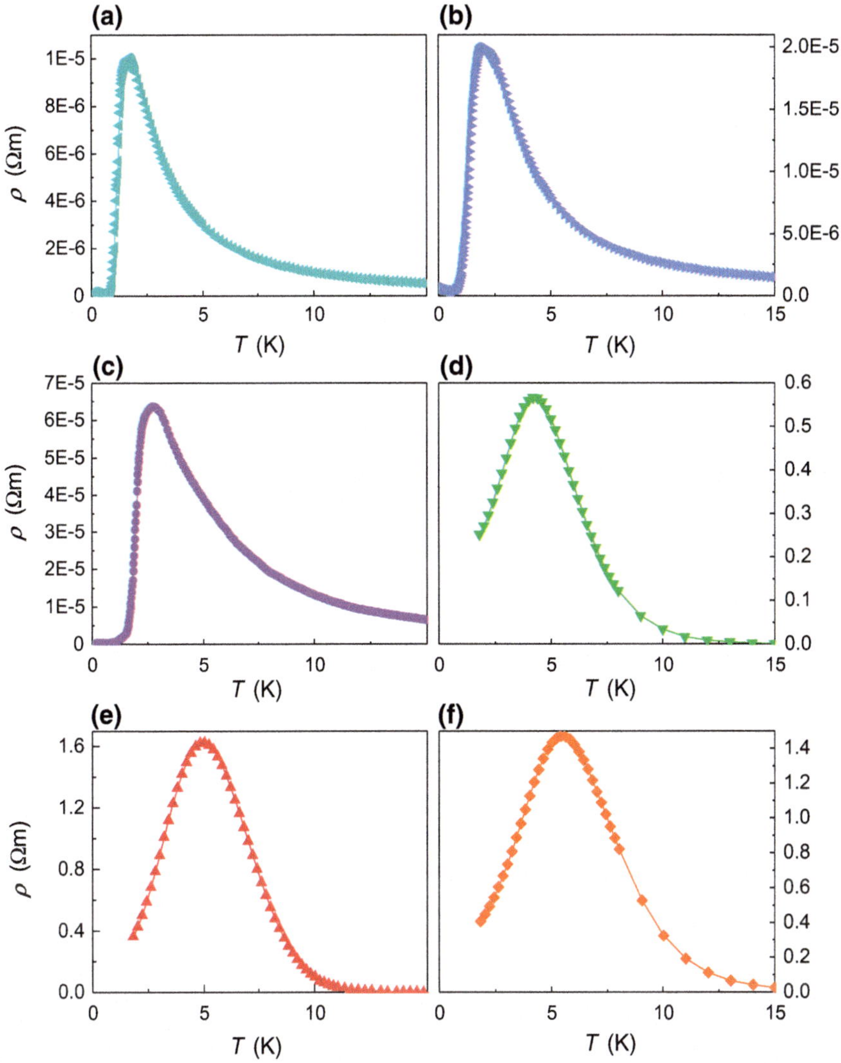

Fig. 6.1 The superconducting transition in $Na_{2-\delta}Mo_6Se_6$ crystals A to F. All data were acquired at $I_{ac} = 10\,\mu A$, $\omega = 1\,Hz$. The small increase in $\rho(T)$ at the lowest temperature regime visible in crystal A and B will be the subject of Chap. 7

Such observations in q1D superconductors are not unusual. For example, it has been shown that carbon nanotube arrays lose their phase coherence at 2–3 T although their Cooper pairs survive fields up to 28 T [10].

From the experimental data, it is confirmed that the ground state of $Na_{2-\delta}Mo_6Se_6$ is superconducting: $\rho(T \rightarrow 0) \sim 0$ and a small Meissner fraction are both observed.

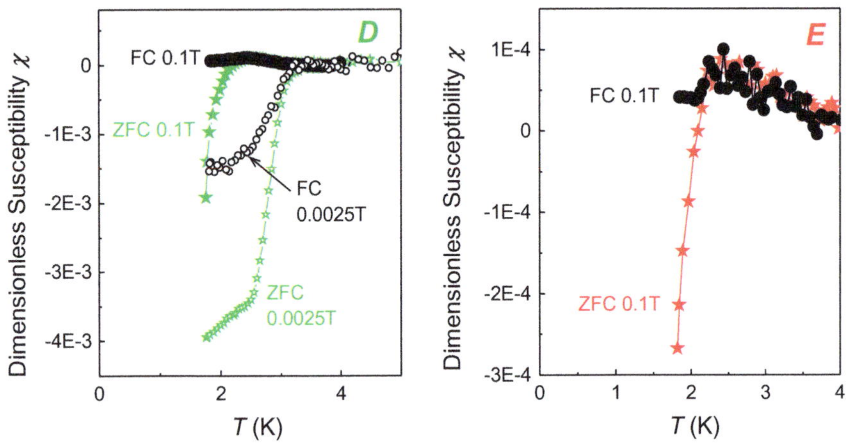

Fig. 6.2 Measurements of the dimensionless susceptibility χ as a function of temperature T indicate an emerging Meissner state in crystals D and E

6.3 The Superconducting Transition

6.3.1 Emergence of 1D Phase-Fluctuating Superconductivity

The 1D crystalline structure of $Na_{2-\delta}Mo_6Se_6$ and the value of $T_{ons} > t_{\perp}^2/t_{\parallel}$ indicate that the superconducting state might be subject to phase fluctuations in its order parameter (Sect. 2.2). Fitting to the phase slip models discussed in Sect. 2.2 was therefore carried out in all crystals (Fig. 6.3). The fits were performed using Eq. 2.24, which includes TAPS and QPS signatures and Eqs. 2.27 and 2.28 which renormalize the TAPS fitting parameter for q1D systems.

In crystals A, B, C, the fit accurately describes the beginning of the superconducting transition but fails at lower temperature (see insets). Instead, a hump develops in the data and is highlighted in the fit by the transition from solid to dashed lines. The physical meaning of this hump will be discussed in the next section.

In crystals D, E, F, the fit follows the transition down to the lowest measured temperature. In systems with higher levels of disorder, the QPS contribution is enhanced [11]. A QPS contribution is expected to be relevant at $T < 0.86T_{ons}$ and causes a tail in the transport measurements leading to $\rho(T \to 0) \neq 0$ (Sect. 2.2.1). The broadening of the peak in crystals D, E, F is therefore attributed to a large contribution of QPS in the phase slip mechanism [5].

In typical 1D nanowire measurements, the TAPS fitting procedure is performed using ξ as a fitting parameter, since the length L of the wire is known [12, 13]. In $Na_{2-\delta}Mo_6Se_6$, L can lie anywhere between ξ and the length of the crystal due to the presence of inhomogeneities and possible macroscopic defects (e.g. cracks). Therefore, the fits were performed using the renormalized expression $Lm/n\xi$ (devel-

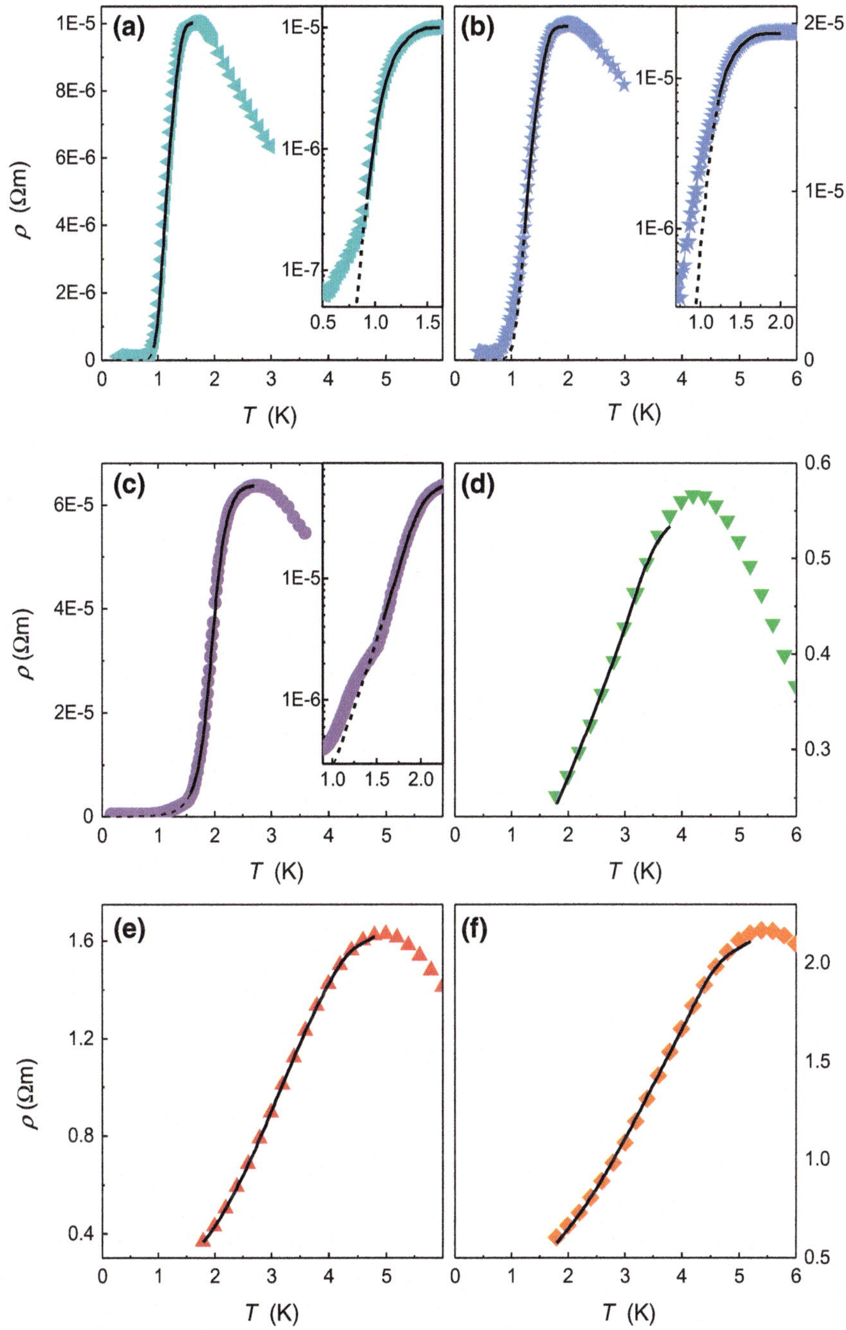

Fig. 6.3 Fits to $\rho(T)$ for 1D phase slips in crystals A to F. **Insets**: Highlights showing the divergence of $\rho(T)$ from the phase slip model in crystals A, B, C

Table 6.1 Phase slip fitting parameters for crystals A to F

Crystal	A_Q	B_Q	$Lm/n\xi$
A	1.99	19400	0.0145
B	1.81	2330	0.0120
C	0.607	44.4	0.0053
D	0.176	0.379	435
E	0.118	0.385	603
F	0.150	0.337	294

oped for q1D systems, Sect. 2.2.3) as a fitting parameter. The values are presented in Table 6.1 together with the fitting parameters A_Q, B_Q for QPS contributions.

In q1D superconductors with a large QPS contribution, A_Q is expected to be of order unity and to decrease with increasing QPS rate [14]: A_Q is larger in crystals A, B, C as expected from their reduced disorder. The value of B_Q is proportional to the effective length of the 1D filaments [14] and should hence fall as the level of disorder rises: this trend is also observed.

Conversely, $Lm/n\xi$ rises by a factor of 10^5 between crystals C and D, which corresponds to the crossing of the mobility edge (Sect. 5.4.4). Several features contribute to this result [5].

First, ξ falls when disorder increases as predicted for dirty superconductors [15].

Second, spatial inhomogeneity developing in the pairing interaction upon crossing the mobility edge strongly contributes to a further reduction of ξ [6, 9, 16, 17].

Third, the electrical transport in $Na_{2-\delta}Mo_6Se_6$ is highly percolative (Sect. 4.4.3) due to disorder. The presence of inhomogeneity creates small islands of superconductivity within the localized material and reduces the number of parallel superconducting filaments n within a typical cross-section of the crystal.

The drastic reduction in $Lm/n\xi$ in crystals D, E, F supports the small Meissner fraction observed in Fig. 6.2, as only limited sections of the crystal are superconducting due to the intense level of disorder.

The emergence of 1D superconductivity in $Na_{2-\delta}Mo_6Se_6$ can now be summarized: TAPS dominate the resistivity immediately below T_{ons} in the less disordered crystals A, B, C and the contribution from QPS is negligible. The influence of QPS rises strongly in crystals D, E, F after crossing the mobility edge [5].

6.3.2 *Establishment of a Coherent Superconducting State*

A hump develops in $\rho(T)$ in crystals A, B, C (inset Fig. 6.3), which in 2D materials is a typical characteristic of a Berezinskii–Kosterlitz–Thouless (BKT) transition [3, 10]. The properties of coupled filamentary materials [18] exhibit similarities to the 2D Coulomb gas, where long-range order is established via a BKT transition. BKT describes a transition from free vortices and antivortices to vortex-antivortex pairs with decreasing temperature [19, 20]. In Josephson-coupled arrays, the BKT

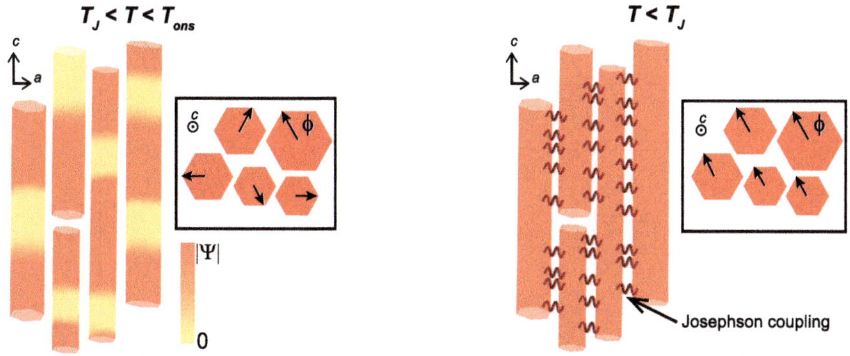

Fig. 6.4 Establishment of phase-coherent superconductivity in $Na_{2-\delta}Mo_6Se_6$. The schematic filaments represent the 1D Na-doped Mo_6Se_6 chains extending along the c axis. The amplitude of the superconducting order parameter $|\Psi|$ is indicated via yellow-red shading, and its phase ϕ only becomes globally coherent via a coupling between the filaments (insets)

transition coincides with the onset of Josephson coupling: $T_{BKT} = T_J$, inducing phase coherence in the system. Although developed for 2D systems, BKT-type transitions have also been identified in Josephson chains [21], carbon nanotube arrays [10], disordered 1D bosonic systems [22] and $Tl_{2-\delta}Mo_6Se_6$ [3]. Furthermore, recent Monte Carlo simulations of Josephson-coupled filamentary superconductors [23] revealed an exponential divergence of the phase correlation length close to the superconducting transition (one of the BKT transition signatures, see below), thus confirming the transition to be of the BKT universality class.

A schematic of a filamentary material such as $Na_{2-\delta}Mo_6Se_6$ is shown in Fig. 6.4 [4]. Above T_J, there is no coupling between the filaments and the superconducting state is dominated by phase fluctuations of the order parameter (yellow shaded areas). At T_J, phase-locking occurs via a BKT-like transition, which takes place in the 2D plane perpendicular to the c axis (insets). To investigate whether this scenario is plausible in $Na_{2-\delta}Mo_6Se_6$, a BKT transition can be tracked via transport measurements.

Several electrical transport characteristics are observable in BKT transitions:

1. The current dependence of the voltage $V(I)$ should exhibit a power-law behavior $V \propto I^{\alpha(T)}$, where α(T) is related to the superfluid density n_s by $\alpha(T) = 1 + \pi n_s \hbar^2 / 4 m_e k_B T$. n_s (and hence α) rises steeply at the BKT transition, with $\alpha(T_{BKT}) = 3$, predicted by Nelson and Kosterlitz [24, 25]. The sharp jump predicted in $\alpha(T)$ at T_{BKT} is substantially broadened by inhomogeneity in real materials [25, 26].
2. The resistance scales exponentially as $R(T) = R_0 \exp\left(-bt^{-1/2}\right)$ over a narrow temperature range above T_{BKT}, where $t = T/T_{BKT} - 1$ and R_0, b are material constants. This was predicted by the Halperin-Nelson model [26–28].
3. Close to T_{BKT}, finite size effects limit the exponential divergence of the correlation length, creating a hump in $\log(R(T))$ [26]. The hump is broadened by current-induced vortex unbinding [29].

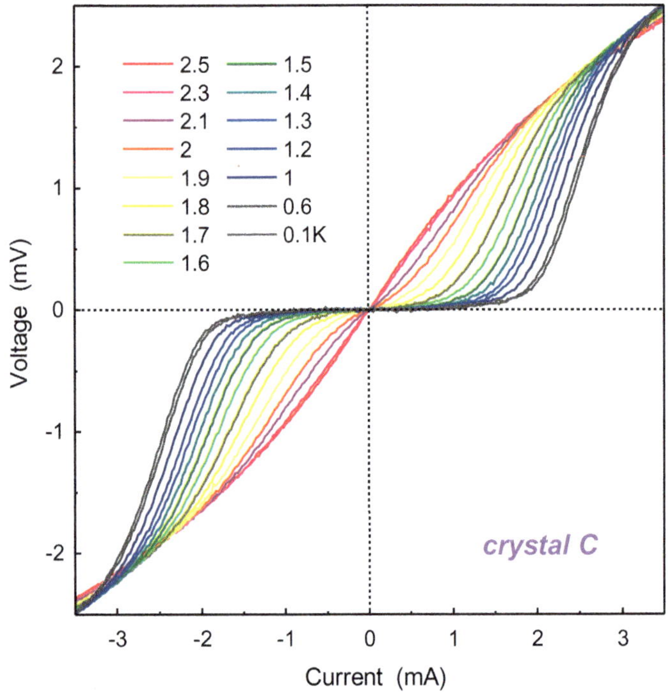

Fig. 6.5 Current dependence of the voltage in crystal C for $T = 0.1$–2.5 K. The data were acquired using 90 μs dc current pulses to prevent sample heating

Together, these characteristics enable a determination of T_{BKT} from transport experiments, as previously demonstrated in a range of 2D [25, 26, 28, 30] and q1D [3, 10, 31] materials.

A current-dependent transport measurement of crystal C has been performed to track the signatures of a BKT transition. The $V(I)$ curves in the temperature range $0.1 < T < 2.5$ K reveal the emergence of a supercurrent for $T < 1.8$ K (Fig. 6.5). Non-Ohmic behavior at high currents and temperature is attributed to sample heating [4].

The $V(I)$ curves display a power-law behavior in the 1–2.5 mA range (Fig. 6.6), where the extracted exponent $\alpha(T)$ exhibits a Nelson-Kosterlitz jump at $\alpha(T) = 3$, establishing $T_{BKT} = 1.72 \pm 0.01$ K [4].

$R(T)$ in crystal C taken at $I = 0.3$ and 0.6 mA follows the exponential expression $R(T) = R_0 \exp\left(-bt^{-1/2}\right)$ over a range ~ 1.8–1.9 K (Fig. 6.7) [4]. By extrapolation, $T_{BKT} = 1.71 \pm 0.03$ K and $T_{BKT} = 1.69 \pm 0.02$ K, in close agreement with the $\alpha = 3$ definition. The determination of the range over which the exponential behavior applies has been theoretically predicted as $X \equiv T(T_{ons} - T_{BKT})/T_{BKT}(T_{ons} - T)$, where X should respect $0 < X - 1 \ll 1$ [28]. In Fig. 6.7 the fitting range is $X = 1.16 - 1.37$, in agreement with theory.

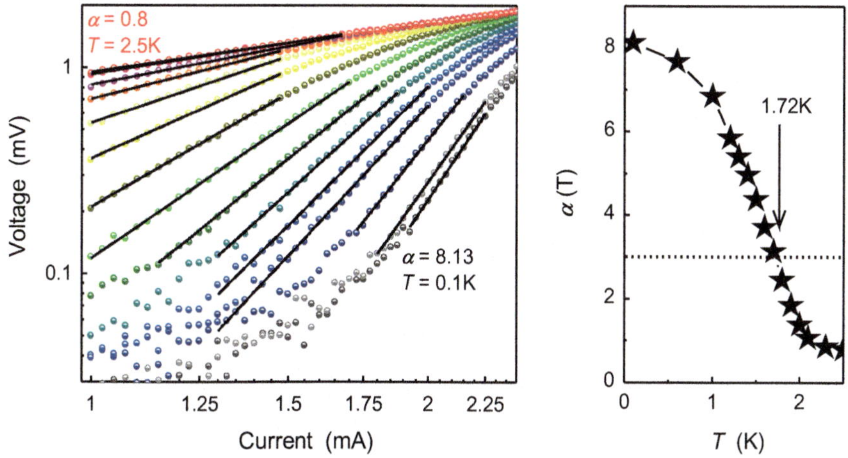

Fig. 6.6 **Left**: Power law $V \propto I^{\alpha(T)}$ fits to the $V(I)$ data depicted in Fig. 6.5 for crystal C. **Right**: Temperature dependence of the exponent α extracted from the power law fits

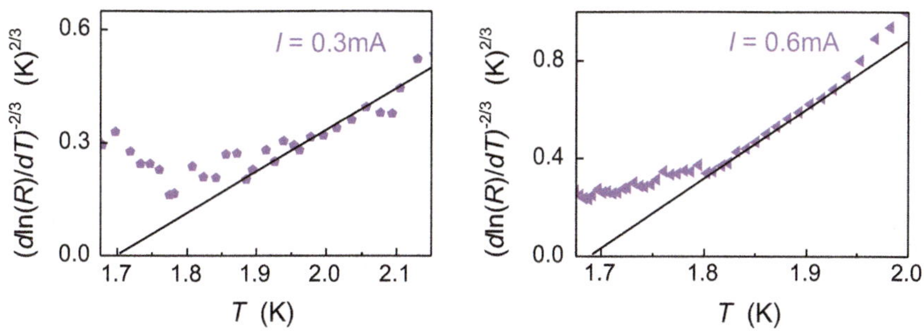

Fig. 6.7 Exponential scaling of $R(T)$ above T_{BKT} for crystal C for applied currents $I = 0.3$ and 0.6 mA. An extrapolation of the linear region to zero (black line) yields $T_{BKT} = 1.71 \pm 0.03$ K and $T_{BKT} = 1.69 \pm 0.02$ K, respectively. However, the fit performed for $I = 0.3$ mA must be considered with caution due to the large noise in the data

The experimentally-determined T_{BKT} exceeds the expected two-particle dimensional crossover temperature $T_J \sim t_{\perp}^2/t_{\parallel} = 1.0$ K. This transverse coupling enhancement is due to two factors [4]. First, any defects (including Na vacancies) strongly reduce the effective $t_{\parallel}$ due to the ease of blocking electron motion along Na-doped $(Mo_6Se_6)_\infty$ filaments. Second, T_J is believed to be enhanced at higher temperatures by increasingly strong e-e interactions [32]. For comparison, $Tl_{2-\delta}Mo_6Se_6$ crystals exhibit a closer correspondence between $T_{BKT} = 4.5$ K [3] and $t_{\perp}^2/t_{\parallel} = 4.4$ K (Chap. 4, Table 4.1) due to their lower Tl deficiency (2.5–5%) [33]. This suggests that the $\sim$ 10% Na vacancy disorder in $Na_{2-\delta}Mo_6Se_6$ has a strong influence on its electronic properties. From the experimental values of T_{BKT}, the Josephson coupling temperature is therefore $T_J = 1.71$ K [4, 5].

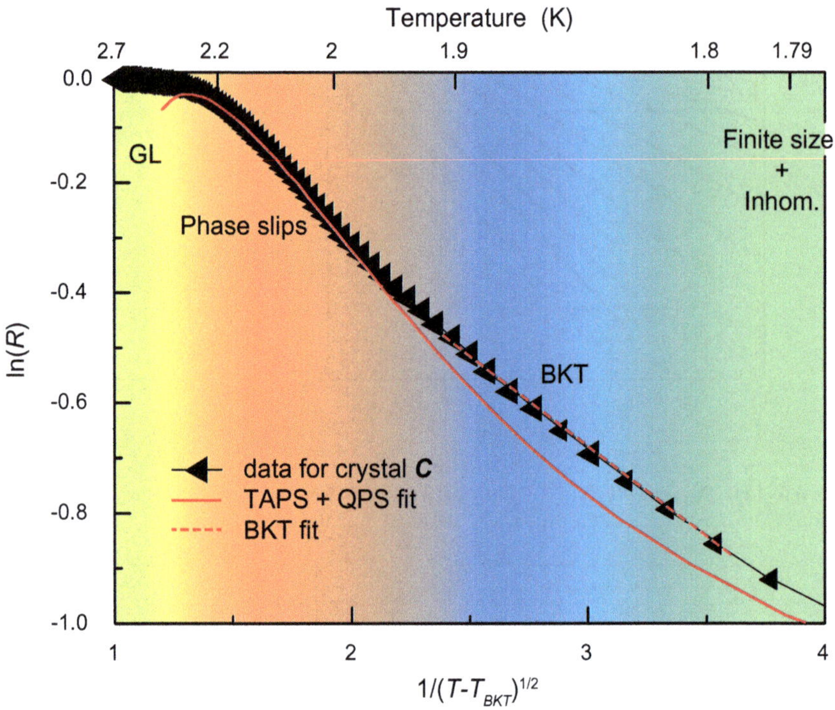

Fig. 6.8 1D–3D crossover from phase slips to BKT-type behavior in crystal C. The resistance is rescaled following the Halperin-Nelson model $\ln R$ versus $1/(T - T_{BKT})^{1/2}$ [27]

The sequential contributions to $R(T < T_{ons})$ in crystal C are summarized in Fig. 6.8 [4]: close to T_{ons}, GL fluctuations of the superconducting order parameter dominate the transport; they are replaced by phase slips as the temperature drops. BKT scaling is only noticeable in a narrow temperature range close to T_{BKT}, limited from above by phase slips and from below by finite size effects and inhomogeneity.

6.4 Pairing Enhancement

The peaks at T_{ons} have been established as the onset of superconductivity and the phase coherence of the superconducting state has been confirmed by the Meissner effect and BKT fits. It is now interesting to explore how their values evolve from crystal A to F. As discussed in Chap. 5, disorder has a clear influence on the electron transport in $Na_{2-\delta}Mo_6Se_6$ and increases from crystal A to F (from the values of ρ(300 K) and VRH T_0). Figure 6.9 illustrates the evolution of T_{ons} as a function of temperature: it rises monotonically from crystal A to F.

T_{ons} as a function of ρ(300 K) in Fig. 6.10 shows a step-like feature between crystals C and D, in accordance with the mobility edge (Sect. 5.4.4). The step observed in

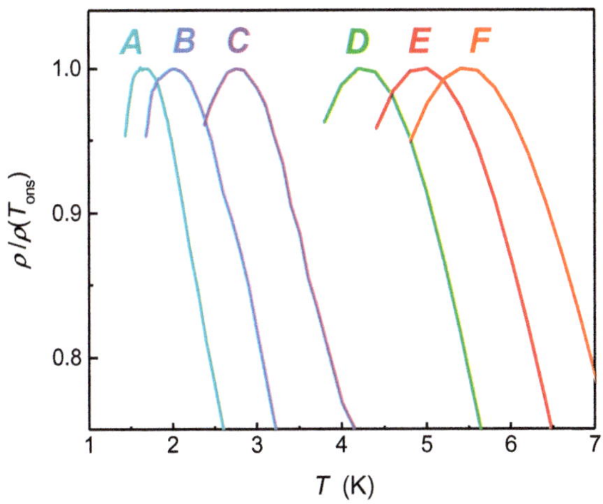

Fig. 6.9 Evolution of $\rho/\rho(T_{ons})$ with temperature T for crystals A to F

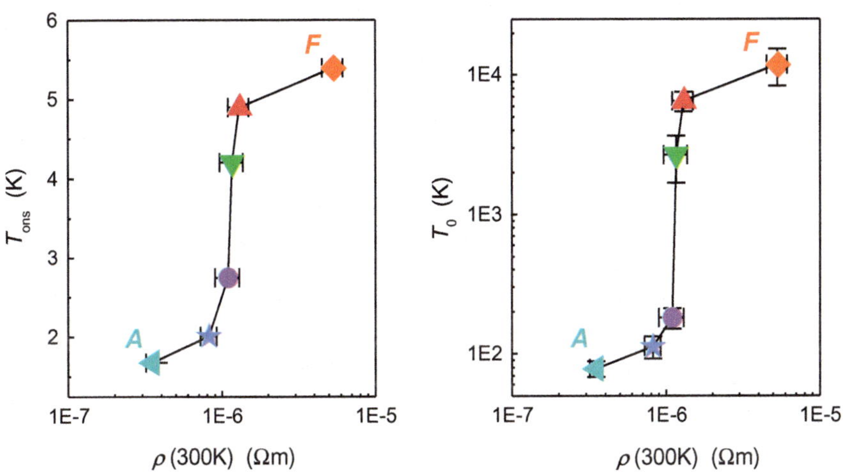

Fig. 6.10 Evolution of the superconducting onset temperature T_{ons} and the localization energy T_0 as a function of ρ(300 K). The right hand panel is similar to Fig. 5.9

Figs. 5.9, 5.14 and 6.10 is therefore due to the presence of the mobility edge between crystals C and D (creating the step-like feature). In consequence, the disorder level in the 6 measured crystals lies very close to the critical disorder, i.e. the mobility edge.

Remarkably, the VRH T_0 displays an identical dependence to ρ(300 K). This suggests that a common physical mechanism drives the enhancement of T_{ons} and T_0. It has been shown that the value of T_0 is related to the level of disorder in the system. From these graphs it is suggested that T_{ons} is also linked to disorder in $Na_{2-\delta}Mo_6Se_6$ [5].

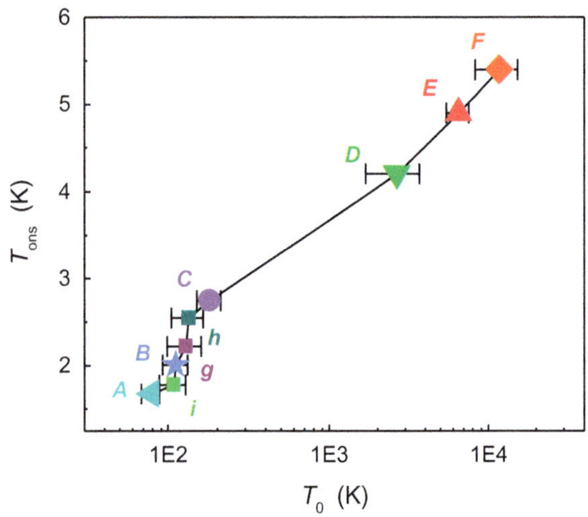

Fig. 6.11 Positive correlation between T_{ons} and T_0

Disorder controls both the superconducting ground state and the insulating upturn in $\rho(T)$ at the intermediate temperature range [5]. Additionally, a plot of T_{ons} as a function of T_0, including all crystals (crystals i, h, g as well) presents a positive correlation between the two parameters (Fig. 6.11). This indicates that the onset temperature for superconducting fluctuations (and hence the pairing energy Δ_0) is enhanced by the presence of localization in $Na_{2-\delta}Mo_6Se_6$ [5].

It is important to consider the distinction between the onset of Cooper pairing and the establishment of long-range phase coherence in $Na_{2-\delta}Mo_6Se_6$ crystals. It has been suggested [16] that although the pairing energy may increase with disorder, the phase stiffness (i.e. the emergence of Josephson coupling) invariably decreases, thus reducing the phase coherence temperature and the true superconducting transition. It is necessary to track the phase stiffness in $Na_{2-\delta}Mo_6Se_6$ and study how it evolves with disorder. Figure 6.12 presents the perpendicular magnetic field dependence of $\rho(T)$ in crystal C and crystal F, i.e. for disorder level below and above the mobility edge. The data were acquired in the PPMS.

In crystal C (below the mobility edge), superconductivity is suppressed by $H_\perp =$ 4 T. In contrast, the superconducting peak at $\rho(T_{ons})$ in crystal F (above the mobility edge) is surprisingly resistant to magnetic fields: a superconducting transition still takes place even at $H_\perp = 14$ T.

For both crystals, the critical magnetic field $H_{c2\perp}(T)$ is extracted as equivalent to $T_{ons}(H_\perp)$. These are plotted below the experimental data: a clear enhancement of $H_{c2\perp}(T)$ is observed between crystal C and F. We can compare the experimental values of $H_{c2\perp}(T)$, which is determined by the coherence length $\xi_\perp$, to the weak-coupling Pauli paramagnetic limit H_P, which depends with the pairing energy Δ_0. If $H_{c2\perp}(T) > H_P$, then the phase stiffness is enhanced.

In singlet-paired superconductors such as $Na_{2-\delta}Mo_6Se_6$, superconducting pairing remains as long as 50% of the Cooper pair spins at E_F prevail antiparallel to the applied magnetic field. A field limit is reached when the energy cost of preserving

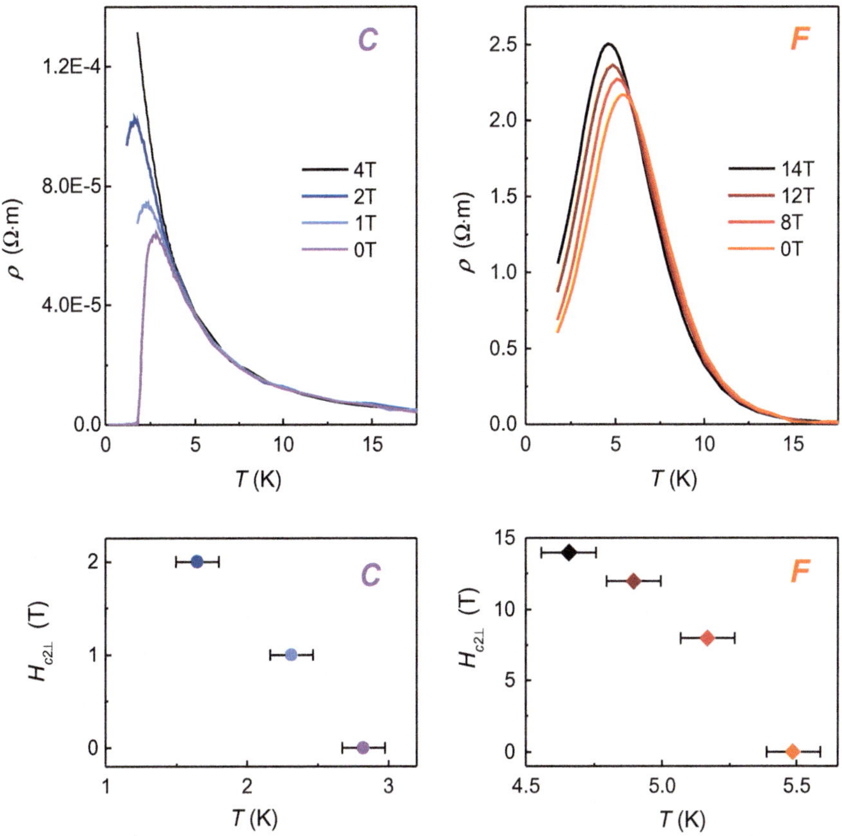

Fig. 6.12 Magnetic field dependence of $\rho(T)$ in crystals C and F in fields perpendicular to the main axis. The extracted critical magnetic fields $H_{c2\perp}(T)$ are obtained from $T_{ons}(H_{\perp})$

the antiparallel spins becomes equal to $\Delta\mathcal{F}_{NS} \equiv \mathcal{F}_N - \mathcal{F}_S$, the difference in free energies between the normal $\mathcal{F}_N$ and the superconducting $\mathcal{F}_S$ phase. This is known as H_P, the weak-coupling Pauli paramagnetic limit or the Clogston–Chandrasekhar limit [34].

In this project, we propose a method to estimate the temperature dependence of H_P [5]. $\Delta\mathcal{F}_{NS} \equiv \mathcal{F}_N - \mathcal{F}_S$ can be expressed in terms of the thermodynamic critical field $H_c(T)$:

$$\frac{H_c^2(T)}{8\pi} = \Delta\mathcal{F}_{NS} \tag{6.1}$$

and for $T = 0$, $\Delta\mathcal{F}_{NS} = 1/2 N_P \Delta_0^2$ where N_P is the superconducting pair density.

$H_c(T)$ can be approximated by:

$$H_c(T) = H_c(0)\left(1 - \left(\frac{T}{T_c}\right)^2\right) \tag{6.2}$$

which allows us to write:

$$\Delta\mathcal{F}_{NS}(T) = \frac{N_P\Delta_0^2}{2}\left(1-\left(\frac{T}{T_c}\right)^2\right)^2 \tag{6.3}$$

The energy cost of a 50% antiparallel spin population is equal to $\mathcal{F}_{AP} = \frac{1}{2}\chi_P H^2$, where the total paramagnetic susceptibility is $\chi_P = 2\mu_B^2 N_\sigma$ and N_σ is the DoS at the Fermi level per spin. At the Pauli limit $H \equiv H_P$ we have $\Delta\mathcal{F}_{NS} = \mathcal{F}_{AP}$, i.e. the energy saved by forming $N_P(T)$ pairs is balanced by the energy cost of maintaining N_σ spins antiparallel to H. Since $N_P(T) = N_\sigma(T) = \frac{1}{2}N(E_F)\left(1-\frac{T}{T_c}\right)$, it leads to:

$$H_P(T) = \frac{\Delta_0}{\sqrt{2}\mu_B}\left(1-\left(\frac{T}{T_c}\right)^2\right) \tag{6.4}$$

For $Na_{2-\delta}Mo_6Se_6$, the value of T_c is set to be equivalent to T_{ons} and the temperature dependence of the Pauli limit $H_P(T)$ can be estimated using the experimental data [5].

Returning to crystal F transport data in Fig. 6.12, $H_{c2\perp} = 14\,\mathrm{T}$ is extracted for $T = 4.6\,\mathrm{K}$. From the above derivation, $H_P = 3\,\mathrm{T}$ is obtained at the same temperature [5]. Hence, in the crystal with disorder above the mobility edge $H_{c2\perp}$ exceeds H_P by a factor > 4. On the other hand, in crystal C (below the mobility edge) $H_{c2\perp} \sim H_P$. An enhancement of $H_{c2\perp}$ signifies that the superfluid phase will survive larger applied magnetic field than the expected Pauli limit. Therefore, phase stiffness is greatly enhanced when crossing the mobility edge towards the localized state.

Another trait of the phase coherence enhancement is observed in the magnetoresistance of crystals C and D at $T = 1.8\,\mathrm{K}$ (Fig. 6.13). Positive magnetoresistance indicates a superconducting state, as an increasing H suppresses progressively the superconducting pair density. In crystal C, superconductivity is suppressed for $H_\perp \approx 4\,\mathrm{T}$ and gives way to a giant negative magnetoresistance due to strong localization in the normal state. Adversely, the magnetoresistance in crystal D stays positive up to the highest field, which indicates a greater phase stiffness in this more disordered crystal. The magnetic field data (Figs. 6.12 and 6.13) suggest that disorder lifts H_{c2}, creating strong electronic correlations which raise the pairing energy [6, 16] above the weak-coupling limit [5].

Finally, the phase coherence temperature $T_J = 1.73\,\mathrm{K}$ in crystal C is higher than the pairing temperature $T_{ons} = 1.68\,\mathrm{K}$ in crystal A, which only seems to develop interchain phase coherence at $T_J \sim 0.95\,\mathrm{K}$ (hump in $\rho(T)$, Fig. 6.3). This supports an enhancement in both pairing and phase coherence by disorder [5]. Accordingly, the offset between T_{ons} and T_J is observed to increase with disorder, ranging from $\sim 0.7\,\mathrm{K}$ in crystal A to $\sim 1.4\,\mathrm{K}$ in crystal E.

From the above mentioned experimental evidences: superconductivity is enhanced by disorder in $Na_{2-\delta}Mo_6Se_6$ [5]. T_{ons}, T_J, and H_{c2} substantially increase from the low disordered crystals (A, B, C) to the more disordered ones (D, E, F). The additional crystals (i, h, g) fit within the trend of T_{ons} enhancement due to Anderson localization

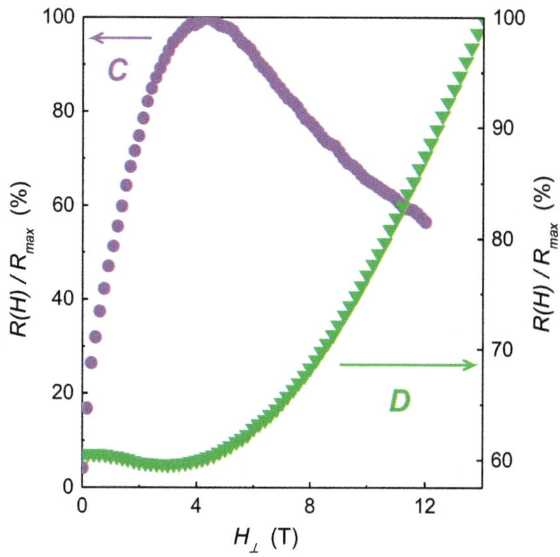

Fig. 6.13 Normalized magnetoresistance $R(H)/R_{max}$ in perpendicular field $H_\perp$ in crystals C and D at $T = 1.8\,\mathrm{K}$

(Fig. 6.11). The mobility edge in the normal state, situated between crystals C and D, is shown to strongly influence the superconducting state (Figs. 6.12 and 6.13). Hence, electron localization in the normal state appears to play a major role in the observed enhancement of superconductivity [5].

6.4.1 Possible Mechanism for the Enhancement of Superconductivity

The enhancement of the superconducting transition temperature has been reported in several superconductors since the 1960s.

First in inhomogeneous aluminum films, an enhancement of T_c by at least a factor of 3 was reported in the 1960s [35]. This effect was attributed to a surface-induced reduction in the phonon frequencies [36]. Later, clustering effects [37] and localized magnetic moments at the grain boundaries [38] were proposed as plausible reasons for the enhancement. Recently, an accurate description of the evolution of T_c in metallic nanoparticles has been achieved using a combination of quantum confinement and mass renormalization due to e-ph scattering [39]. For such mechanism, no report of a large enhancement in the pairing energy was found in the literature, which contrasts with the T_J and H_{c2} enhancement observed in $Na_{2-\delta}Mo_6Se_6$.

Second in low-dimensional superconductors, impurities have been reported to increase T_J [40] and disorder to enhance the transverse e-ph coupling [41]. These effects are only related to the transverse coupling and are not relevant to the increase in T_{ons}. Alternatively, Anderson U impurities have been proposed to locally increase e-ph coupling [42], but a Na vacancy cannot be considered a U impurity since it merely leaves a small charge deficit to be screened on the MoSe chains, without any local U enhancement [43].

Third in amorphous Bi films, superconductivity may be induced with $T_c >$ 6 K [44]. This is explained by an enhancement in the DoS at the Fermi level within amorphous Bi compared to its crystalline form. Yet, increasing disorder in amorphous Bi monotonically suppresses T_c [45].

Finally in the Chevrel phase $PbMo_6S_8$, T_c rises up to 15 K and H_{c2} exceeds 80 T. These properties are due to high densities of states and multiple bands crossing the Fermi level, respectively [46]. The electronic structure of the $M_{2-\delta}Mo_6Se_6$ family is rather different: $N(E_F)$ is more than an order of magnitude lower and a single 1D band crosses E_F.

This brief literature review indicates that these reported mechanisms cannot be responsible for the enhancement of superconductivity in $Na_{2-\delta}Mo_6Se_6$. Instead, a positive relationship between disorder-induced electron localization and superconductivity is consistently observed in our data.

A separate mechanism for the enhancement of superconductivity in strongly disordered materials suggests that in the absence of Coulomb repulsion, T_c actually rises with increasing normal state resistance [47]. In highly disordered systems, the electron wavefunctions near the localization mobility edge E_c become multifractalized: they are partially localized and yet still extend through the system [48]. The uniform and spatially-distributed electron waves are replaced by multifractalized electron eigenfunctions, which only occupy a fraction of the volume within their correlation radii [7]. This leads to a mosaic-like spatial distribution for each electronic wavefunction, composed of finite fractal elements. An illustration from Schenk et al. [49] of the extended ($E_F << E_c$), localized ($E_F >> E_c$), and multifractalized ($E_F \sim E_c$) wave function probabilities in a 3D system is presented in Fig. 6.14: the multifractalized state extends nearly everywhere while exhibiting localized regions with higher electron site probability.

The formation of local peaks in the multifractal distribution enhances e-e correlations: if a superconducting instability is present, the pairing energy and the critical temperature may be increased [6, 7, 47, 50]. This is presented in a simplified graph in Fig. 6.15 [47]: T_c is enhanced when the Fermi energy is crossing the mobility edge due to increasing disorder. However, for a too large level of disorder far into the localized state, T_c is expected to decrease again.

Several theoretical and simulation studies have investigated the multifractality of the electronic wavefunction [49, 51–53]. On the experimental side, a multifractal electronic state has previously been observed in strongly localized 1D metals [54], but no associated enhancement of superconductivity has yet been reported.

$Na_{2-\delta}Mo_6Se_6$ has been shown to exhibit Anderson localization in the normal state prior to the emergence of superconductivity [5].

Furthermore, the observation of a mobility edge between crystals C and D indicates that the disorder level in the measured crystals lies where multifractality of the electron wavefunction is expected to take place. In order to know if this mechanism

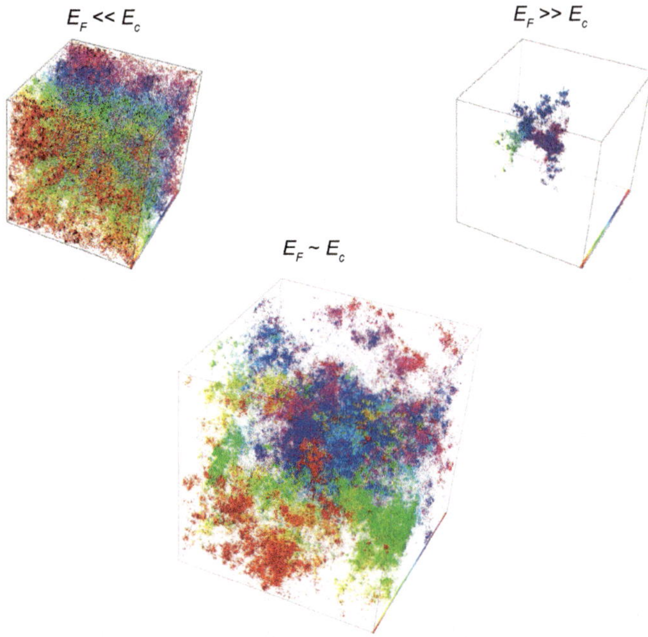

Fig. 6.14 Illustration of the extended ($E_F << E_c$), localized ($E_F >> E_c$), and multifractalized ($E_F \sim E_c$) wave function probabilities in a 3D system subject to Anderson localization. The colors are a guide to the eye of the different slices along the axis going into the page. Adapted from Schenk et al. [49]

Fig. 6.15 Evolution of T_c as a function of the Fermi energy position E_F relative to the mobility edge. Adapted from Kravtsov [47]

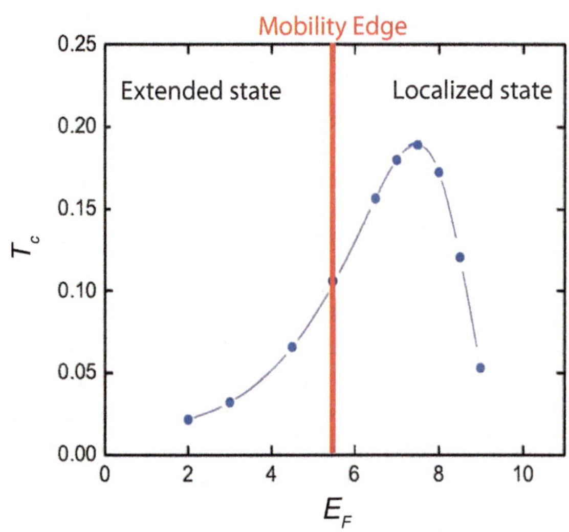

can be responsible for the enhancement of T_{ons}, T_J and H_{c2}, it is necessary to prove that long-range Coulomb repulsion is screened or negligible.

The first signs of screened Coulomb repulsion is found in the calculated DoS (Fig. 4.9): the peak present in $N(E)$ below the Fermi level suggests that the background dielectric constant is large, leading to screened Coulomb repulsion [5]. Moreover, the very low DoS at the Fermi level indicates that if Coulomb repulsion was present in the material, it would be weak [5].

Secondly, the high temperature regime presents characteristics of a LL with attractive e-e interaction (Fig. 5.4). This implies that Coulomb repulsion is weak or negligible [5].

Thirdly, any opening of a soft Coulomb gap at the Fermi level would be visible in $\rho(T)$ at the intermediate temperature regime via Efros–Shklovskii VRH with dimension $d = 1$. The fits to the VRH model yield $d > 1$ (Table 5.2), supporting that no gap opens in $Na_{2-\delta}Mo_6Se_6$ [5].

Finally, the magnetoresistance $R(H_\perp)$ in crystal D at $T = 150$ K and C at $T = 1.8$ K (Figs. 5.12 and 6.13) support the scenario of Anderson localization with no Coulomb gap: the strong negative $R(H)$ behavior is due to delocalized electrons; the presence of a gap would yield a positive behavior [5].

These observations support the scenario of a multifractal enhancement of superconductivity in $Na_{2-\delta}Mo_6Se_6$ with increasing disorder [5]. Quantitative predictions for multifractal enhancement have been made only in 2D and 3D superconductors [50]. Nevertheless, an adaptation of these existing models to the q1D geometry of $Na_{2-\delta}Mo_6Se_6$ has been discussed in A.P. Petrović et al. [5]. There, it is demonstrated how the model developed by Burmistrov et al. [50] is closely compatible with the experimental data in $Na_{2-\delta}Mo_6Se_6$. Fits to the enhancement of T_{ons} provide values of the multifractal exponent comparable to the predicted values for 2D and 3D systems, supporting this scenario for a q1D material [5]. This analysis is beyond the scope of this thesis although as discussed in the Supplementary of Ref. [5], the results provide optimism for further studies in this direction.

6.5 Concluding Remarks

The superconducting transition in $Na_{2-\delta}Mo_6Se_6$ was studied in detail via transport and magnetization measurements. The data analysis indicates that first, superconductivity is established within single 1D filaments and is dominated by phase slips, creating phase fluctuations of the order parameter. With decreasing temperature, Josephson coupling takes place between the filaments, establishing phase coherence within the system. Therefore, the superconducting state undergoes a 1D–3D crossover via Josephson coupling between nanofilaments.

The temperature of the coupling is well described by the BKT model, suggesting that a BKT-like transition takes place in this q1D system within the 2D-plane perpendicular to the chains. This was previously observed in other q1D systems and theoretically confirmed via Monte-Carlo simulations.

The presence of disorder in the studied q1D system showed a striking effect in the superconducting ground state: an enhancement of both the superconducting onset temperature T_{ons} and the phase coherence with increasing disorder. This was observed in transport and magneto-transport data. The data analysis indicated that electron localization in the normal state and the suppression of the Coulomb repulsion may be responsible for this effect, as it was recently theoretically proposed.

References

1. J.C. Armici, M. Decroux, Ø. Fischer, M. Potel, R. Chevrel, M. Sergent, A new pseudo-one-dimensional superconductor: Tl2Mo6Se6. Solid State Commun. **33**(254), 607–611 (1980)
2. A.P. Petrović, R. Lortz, G. Santi, M. Decroux, Phonon mode spectroscopy, electron-phonon coupling, and the metal-insulator transition in quasi-one-dimensional M2Mo6Se6. Phys. Rev. B **82**(23), 235128 (2010)
3. B. Bergk, A.P. Petrović, Z. Wang, Y. Wang, D. Salloum, P. Gougeon, M. Potel, R. Lortz, Superconducting transitions of intrinsic arrays of weakly coupled one-dimensional superconducting chains: the case of the extreme quasi-1D superconductor Tl2Mo6Se6. New J. Phys. **13**(10), 103018 (2011)
4. D. Ansermet, A.P. Petrović, S. He, D. Chernyshov, M. Hoesch, D. Salloum, P. Gougeon, M. Potel, L. Boeri, O.K. Andersen, C. Panagopoulos, Reentrant phase coherence in superconducting nanowire composites. ACS Nano **10**(1), 515–523 (2016)
5. A.P. Petrović, D. Ansermet, D. Chernyshov, M. Hoesch, D. Salloum, P. Gougeon, M. Potel, L. Boeri, C. Panagopoulos, A disorder-enhanced quasi-one-dimensional superconductor. Nat. Commun. **7**, 12262 (2016)
6. M. Feigel'man, L. Ioffe, V. Kravtsov, E. Yuzbashyan, Eigenfunction fractality and pseudogap state near the superconductor-insulator transition. Phys. Rev. Lett. **98**(2), 027001 (2007)
7. M.V. Feigel'man, L.B. Ioffe, V.E. Kravtsov, E. Cuevas, Fractal superconductivity near localization threshold. Ann. Phys. **325**(7), 1390–1478 (2010)
8. A. Ghosal, M. Randeria, N. Trivedi, Role of spatial amplitude fluctuations in highly disordered s-wave superconductors. **12**(1), 4 (1998)
9. Y. Dubi, Y. Meir, Y. Avishai, Nature of the superconductor-insulator transition in disordered superconductors. Nature **449**(7164), 876–80 (2007)
10. Z. Wang, W. Shi, R. Lortz, P. Sheng, Superconductivity in 4-angstrom carbon nanotubes - a short review. Nanoscale **4**(1), 21–41 (2012)
11. C. Cirillo, M. Trezza, F. Chiarella, A. Vecchione, V.P. Bondarenko, S.L. Prischepa, C. Attanasio, Quantum phase slips in superconducting Nb nanowire networks deposited on self-assembled Si templates. Appl. Phys. Lett. **101**(17), 172601 (2012)
12. C.N. Lau, N. Markovic, M. Bockrath, A. Bezryadin, M. Tinkham, Quantum phase slips in superconducting nanowires. Phys. Rev. Lett. **87**(21), 217003 (2001)
13. A. Rogachev, A. Bollinger, A. Bezryadin, Influence of high magnetic fields on the superconducting transition of one-dimensional Nb and MoGe nanowires. Phys. Rev. Lett. **94**(1), 017004 (2005)
14. D.S. Golubev, A.D. Zaikin, Quantum tunneling of the order parameter in superconducting nanowires. Phys. Rev. B **64**(1), 014504 (2001)
15. M.V. Sadovskii, Superconductivity and localization. Phys. Rep. **282**(5), 225–348 (1997)
16. A. Ghosal, M. Randeria, N. Trivedi, Inhomogeneous pairing in highly disordered s-wave superconductors. Phys. Rev. B **65**, 014501 (2001)
17. K. Bouadim, Y.L. Loh, M. Randeria, N. Trivedi, Single- and two-particle energy gaps across the disorder-driven superconductor-insulator transition. Nat. Phys. **7**(11), 884–889 (2011)

18. H.J. Schulz, C. Bourbonnais, Quantum fluctuations in quasi-one-dimensional superconductors. Phys. Rev. B **27**(9), 5856–5859 (1983)
19. V.L. Berezinskii, Destruction of long-range order in one-dimensional and two-dimensional systems having a continuous symmetry group I. Classical systems. Sov. Phys. JETP **32**(3), 2–9 (1971)
20. J.M. Kosterlitz, D.J. Thouless, Long range order and metastability in 2D solids ans superfluids. J. Phys. C **5**, L124–L126 (1972)
21. R.M. Bradley, S. Doniach, Quantum fluctuations in chains of Josephson junctions. Phys. Rev. B **30**(3), 1138 (1984)
22. E. Altman, Y. Kafri, A. Polkovnikov, G. Refael, Phase transition in a system of one-dimensional bosons with strong disorder. Phys. Rev. Lett. **93**(15), 1–4 (2004)
23. M.Y. Sun, Z.L. Hou, T. Zhang, Z. Wang, W. Shi, R. Lortz, P. Sheng, Dimensional crossover transition in a system of weakly coupled superconducting nanowires. New J. Phys. **14**(10), 103018 (2012)
24. D.R. Nelson, J.M. Kosterlitz, Universal jump in the superfluid density of two-dimensional superfluids. Phys. Rev. Lett. **39**(19), 1201–1205 (1977)
25. A.M. Kadin, K. Epstein, A.M. Goldman, Renormalization and the Kosterlitz-Thouless transition in a two-dimensional superconductor. Phys. Rev. B **27**(11), 6691–6702 (1983)
26. L. Benfatto, C. Castellani, T. Giamarchi, Broadening of the Berezinskii-Kosterlitz-Thouless superconducting transition by inhomogeneity and finite-size effects. Phys. Rev. B **80**(21), 214506 (2009)
27. B.I. Halperin, D.R. Nelson, Resistive transition in superconducting films. J. Low Temp. Phys. **36**(5/6), 599–616 (1979)
28. P. Minnhagen, The two-dimensional coulomb gas, vortex unbinding, and superfluid-superconducting films. Rev. Mod. Phys. **59**(4), 1001–1066 (1987)
29. N. Coton, M.V. Ramallo, F. Vidal, Effects of critical temperature inhomogeneities on the voltage-current characteristics of a planar superconductor near the Berezinskii-Kosterlitz-Thouless transition. Supercond. Sci. Technol. **24**(8), 085013 (2011)
30. N. Reyren, S. Thiel, A.D. Caviglia, L.F. Kourkoutis, G. Hammerl, C. Richter, C.W. Schneider, T. Kopp, A.-S. Rüetschi, D. Jaccard, M. Gabay, D.A. Muller, J.-M. Triscone, J. Mannhart, Superconducting interfaces between insulating oxides. Science (New York) **317**(5842), 1196–1199 (2007)
31. M. He, C.H. Wong, Di. Shi, P.L. Tse, E.-W. Scheidt, G. Eickerling, W. Scherer, P. Sheng, R. Lortz, 1D to 3D dimensional crossover in the superconducting transition of the quasi-one-dimensional carbide superconductor Sc3CoC4. J. Phys. Condens. Matter **27**(7), 075702 (2015)
32. T. Giamarchi, Theoretical framework for quasi-one dimensional systems. Chem. Rev. **104**(11), 5037–5056 (2004)
33. R. Brusetti, A. Briggs, O. Laborde, M. Potel, P. Gougeon, Superconducting and dielectric instabilities in Tl2Mo6Se6: unusual transport properties and unsaturating critical field. Phys. Rev. B **49**(13), 8931–8943 (1994)
34. A.M. Clogston, Upper limit for the critical field in hard superconductors. Phys. Rev. Lett. **9**(6), 5–6 (1962)
35. B. Abeles, R.W. Cohen, G.W. Cullen, Enhancement of superconductivity in metal films. Phys. Rev. Lett. **17**(12), 632–634 (1966)
36. J.J. Hauser, Enhancement of superconductivity in aluminum films. Phys. Rev. B **3**(5), 1611–1616 (1971)
37. B. Cao, C.M. Neal, A.K. Starace, Y.N. Ovchinnikov, V.Z. Kresin, M.F. Jarrold, Evidence for high Tc superconducting transitions in isolated Al45- and Al47- nanoclusters. J. Supercond. Nov. Magn. **21**(3), 163–166 (2008)
38. N. Bachar, S. Lerer, S. Hacohen-Gourgy, B. Almog, G. Deutscher, Kondo-like behavior near the metal-to-insulator transition of nanoscale granular aluminum. Phys. Rev. B Condens. Matter Mater. Phys. **87**(21), 2–5 (2013)
39. M.D. Croitoru, A.A. Shanenko, A. Vagov, M.V. Milošević, V.M. Axt, F.M. Peeters, Phonon limited superconducting correlations in metallic nanograins. Sci. Rep. **5**, 16515 (2015)

40. K.B. Efetov, A.I. Larkin, Fluctuation effect on the transition temperature in quasi-one-dimensional superconductors. Sov. Phys. JETP **39**, 1129–1134 (1974)
41. B. Keck, A. Schmid, Superconductivity and electron-phonon interaction in impure simple metals. J. Low Temp. Phys. **24**(5–6), 611–629 (1976)
42. I. Martin, P. Phillips, Local pairing at U impurities in BCS superconductors can enhance Tc. Phys. Rev. B **56**(22), 14650–14654 (1997). ISSN 0163-1829
43. P.W. Anderson, Theory of localized magnetic states in metals. J. Appl. Phys. **37**(3), 1194 (1961). ISSN 00218979
44. W. Buckel, Elektronenbeugungs-aufnahmen von dünnen metallschichten bei tiefen temperaturen. Z. für Phys. **138**(2), 136–150 (1954)
45. D.B. Haviland, Y. Liu, A.M. Goldman, Onset of superconductivity in the two-dimensional limit. Phys. Rev. Lett. **62**(18), 2180–2183 (1989)
46. A.P. Petrović, R. Lortz, G. Santi, C. Berthod, C. Dubois, M. Decroux, A. Demuer, A.B. Antunes, A. Paré, D. Salloum, P. Gougeon, M. Potel, Ø. Fischer, Multiband superconductivity in the chevrel phases SnMo6S8 and PbMo6S8. Phys. Rev. Lett. **106**(1), 017003 (2011)
47. V.E. Kravtsov, Wonderful life at weak coulomb interaction: increasing of superconducting/superfluid transition temperature by disorder. J. Phys. Conf. Ser. **376**, 012003 (2012)
48. H. Aoki, Critical behaviour of extended states in disordered systems. J. Phys. C Solid State Phys. **205**(16), L205–L208 (1983)
49. O. Schenk, M. Bollhöfer, R.A. Römer, On large-scale diagonalization techniques for the Anderson model of localization. SIAM Rev. **50**(1), 91–112 (2008)
50. I.S. Burmistrov, I.V. Gornyi, A.D. Mirlin, Enhancement of the critical temperature of superconductors by Anderson localization. Phys. Rev. Lett. **108**(1), 017002 (2012)
51. W. Pook, M. Janssen, Multifractality and scaling in disordered mesoscopic systems. Z. für Phys. B Condens. Matter **82**(2), 295–298 (1991)
52. M. Janssen, Multifractal analysis of broadly-distributed observables at criticality. Int. J. Mod. Phys. B **8**(8), 943–984 (1994)
53. I.M. Khaymovich, J.V. Koski, O.-P. Saira, V.E. Kravtsov, J.P. Pekola, Multifractality of random eigenfunctions and generalization of Jarzynski equality. Nat. Commun. **6**, 7010 (2015)
54. M. Schreiber, H. Grussbach, Multifractal wave functions at the Anderson transition. Phys. Rev. Lett. **67**(5), 607–610 (1991)

Chapter 7
Reentrant Phase Coherence by Josephson Coupling

7.1 Introduction

In low-dimensional inhomogeneous systems, Josephson coupling can be a necessary step towards phase-coherent superconductivity. The interplay of Josephson coupling in establishing coherence in granular superconductors has been widely studied (see, for example, reviews from Gantmakher [1] or Beloborodov et al. [2]). In addition, other systems have shown the importance of intrinsic Josephson junctions in their structures. For example, Josephson coupling can be responsible for long-range order in systems including 2D cuprates [3], pnictides [4] and q1D filamentary materials [5]. Here, the weak links form naturally between atomic planes or chains within highly anisotropic crystal structures.

The similarities in the Josephson coupling between a 2D granular film and a q1D crystal can be understood via a schematic (Fig. 7.1). In a granular film, Cooper pairs hop between two grains, similar to hopping between two filaments in a q1D material. The theories developed for granular systems may therefore be adapted to filamentary materials [6].

It was shown in Chap. 6 that a coherent superconducting state is established in $Na_{2-\delta}Mo_6Se_6$ in crystals A, B, C ($R(T \to 0) \sim 0$, Fig. 6.1). These crystals were measured down to mK temperatures and the stability of the superconducting ground state was investigated in depth in crystal C.

This chapter presents the results of this study: not only does Josephson coupling play an eminent role in establishing a superconducting state, it also controls the stability of this state as a function of temperature T, magnetic field H and current I. This work shows that a coherent superconducting state develops at non-zero (T, H, I), which constitutes reentrant superconductivity. Additionally, it is shown how such behavior in filamentary materials can be predicted by the Josephson energy. This work was published in 2016 in *ACS Nano* [6].

D. Ansermet, *Emergent Superconductivity in Low Dimensions*, Springer Theses,
https://doi.org/10.1007/978-981-13-2941-8_7

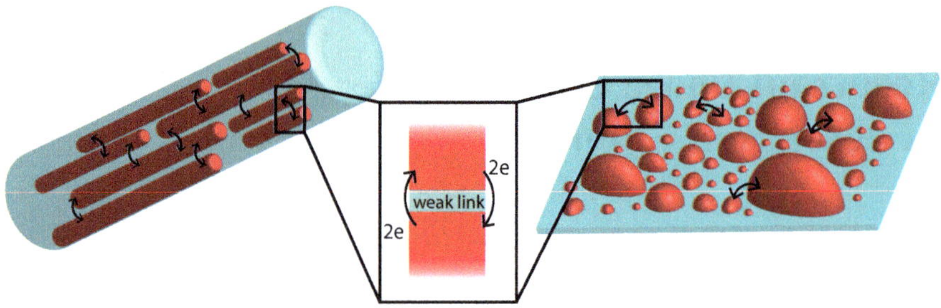

Fig. 7.1 Schematic of Josephson coupling in q1D filamentary (left) and 2D granular (right) systems

7.2 Reentrant Superconductivity: A Short Overview

A reentrant behavior describes the development or stabilization of superconductivity upon raising T, H or I. This phenomenon has been predicted and observed in various systems [7–32]. The majority of technological applications require superconductors to operate at non-zero (T, H, I), which implies that reentrance is a highly desirable feature.

7.2.1 Reentrant Superconductivity by Josephson Coupling

The very first experimental observation of reentrance was made by Lin et al. in 1984 [7] (Fig. 7.2a), following two independent predictions of the effects of charging energy on the critical temperature in granular superconductors by E. Simanek in 1979 [8] and K.B. Efetov in 1980 [9]. Lin et al. analyzed granular $BaPb_{0.75}Bi_{0.25}O_3$ and attributed the observed reentrance to the decoupling of the superconducting grains at low temperature, which was associated with the reduction of charge carriers in the grain boundaries. These experimental results motivated deeper analysis of the mechanism controlling reentrance: in 1985, Simanek [10] discussed the interplay between the Josephson energy E_J and the Coulomb repulsion U as the main factor mediating the reentrance in granular systems (Fig. 7.2b). Later, Belevtsev et al. [11] reported a reentrant behavior in ultrathin amorphous films of bismuth (Fig. 7.2c) and explained it through the expression of E_J.

Belevtsev et al. analyzed the reentrant behavior through $E_J(T, I)$ for granular materials. In their model, they considered a system composed of nearly identical grains separated by insulating weak links. Phase coherence develops between the grains following a similar expression to Eq. 2.33:

$$E_J = \frac{\pi\hbar}{4e^2 R_t}\Delta(T)\tanh\left(\frac{\Delta(T)}{2kT}\right) \tag{7.1}$$

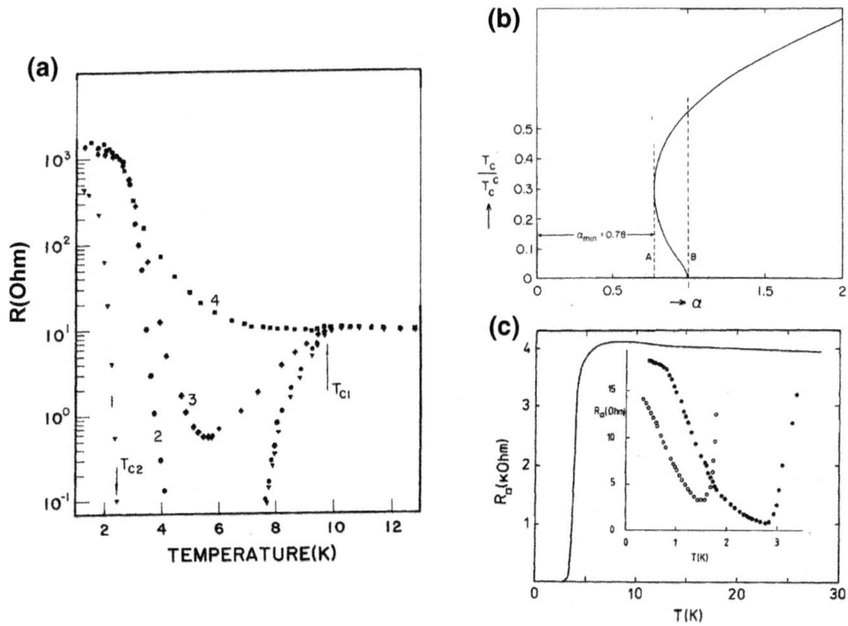

Fig. 7.2 Reentrant superconductivity in granular materials. **a** Reentrance in $R(T)$ experimental data in $BaPb_{0.75}Bi_{0.25}O_3$ for different applied currents and magnetic fields: (1) 11 μA, 0.5 G; (2) 72 μA, 0.5 G; (3) 94 μA, 2.0 G; 72 μA, 216 G. From Lin et al. [7]. **b** Reentrance in the theoretical phase-ordering temperature ratio T_c/T_c^c plotted as a function of parameter $\alpha = zE_J/U$. T_c^c is the critical temperature when U is zero. The dashed lines A and B indicate the reentrant portions of the phase diagram. From Simanek [10]. **c** Reentrance in sheet $R(T)$ in ultrathin amorphous films of bismuth. The inset highlight the tails of the superconducting transition for H = 0 kOe (full circles) and 47 kOe (open circles). From Belevtsev et al. [11]

where R_t is the tunneling resistance between the grains expressed as:

$$R_t^{-1} = \frac{e^2}{\hbar}\frac{\hbar\omega_t}{E_a} \tag{7.2}$$

The activation energy E_a is the characteristic energy scale describing the quantized level mismatch and typical capacitive charging energy $2e^2/C$ of each superconducting grain [9, 33]; ω_t is the frequency of electron jumps between grains, which controls the probability P of tunneling through the weak link:

$$P \sim \omega_t^2 \sim A \exp\left[-\frac{2s}{\hbar}(2m_e\varphi_0)^{1/2}\right] \cdot \exp\left[-\frac{E_a}{k_B T}\right] \tag{7.3}$$

where A is a constant, s the barrier thickness, m_e the electron mass, and $\varphi_0 = \Phi - eV$ the effective barrier height.

This model showed that for inhomogeneous materials, reentrance may occur if $E_J(T, I)$ rises with respect to the thermal energy (or the Coulomb energy in a granular material) [8–13]. A rise in E_J facilitates Cooper pair transfer between the regions: once E_J becomes sufficiently large, phase coherence is established. However, Belevtsev et al. did not explain the observed reentrant behavior with applied magnetic field. Nevertheless, this approach is the most advanced theoretical work developed for reentrant superconductivity by Josephson coupling in inhomogeneous materials.

7.2.2 Reentrant Superconductivity by Other Mechanisms

Several other mechanisms have been proposed to induce a reentrant superconducting state.

Jaccarino–Peter effect: In some ferromagnetic materials, superconductivity can be induced by a large applied magnetic field through the compensation of the exchange field between the magnetic moments and the conduction electrons. This reentrance by field compensation is known as the Jaccarino–Peter effect [14] and has been observed in rare-earth compounds [15], Chevrel phases [16] (Fig. 7.3) and 2D organic insulators [17].

FFLO phase: In superconductors under large applied magnetic field, a Fulde–Ferrell–Larkin–Ovchinnikov (FFLO) state [18, 19] can emerge from Cooper pairs with non-zero total momentum and periodically modulated order parameter.

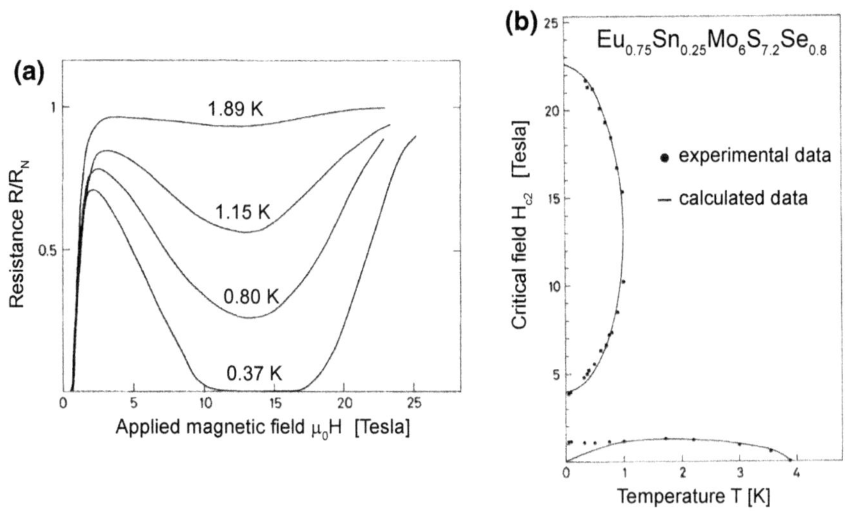

Fig. 7.3 Jaccarino–Peter effect inducing reentrant superconductivity in $Eu_{0.75}Sn_{0.25}Mo_6S_{7.2}Se_{0.8}$. **a** Normalized resistance R/R_N versus field ($H \sim 25\,T$) at various temperatures. **b** Upper critical field H_{c2} versus temperature. From Meul et al. [16]

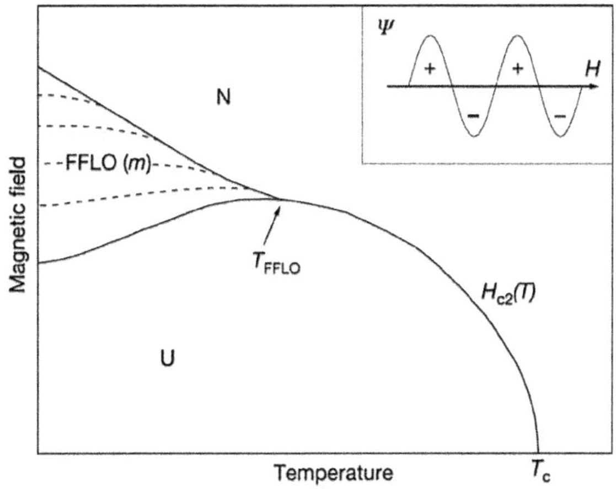

Fig. 7.4 Qualitative phase diagram of the emergence of a FFLO state. Depicted are the uniform superconducting (U), normal (N) and FFLO phases. The inset shows the spatial modulation of the order parameter Ψ along the applied field H within the non-uniform FFLO phase. From Radovan et al. [23]

For a FFLO phase to take place, the Maki parameter $\alpha = \sqrt{2}H_{orb}/H_P$ [34] must be > 1.6, where H_{orb} is the orbital limiting field (due to the overlap of field-induced vortices), typically estimated from the Werthamer–Helfand–Hohenberg formula $H_{orb}(T = 0) = 0.7T_c|\mathrm{d}B_{c2}/\mathrm{d}T|_{T_c}$ [35]. This reentrance mechanism has been reported in low-dimensional Bechgaard salts [20–22], heavy fermion materials [23, 24], and superconductor/ferromagnet layered systems [25], to cite only a few. The typical emergence of a FFLO state is presented in Fig. 7.4.

Magnetic impurities: Magnetic impurities in nanowires and films have been predicted [27] and observed [28] to induce reentrance for a narrow (T, H, I) parameter range. A reduction in the exchange scattering rate by impurity polarization is believed to cause reentrance. This effect is induced by applied magnetic field $H \neq 0$ and large applied current $I \geq I_c$.

Vortex pinning: A reentrant behavior has been reported by Cordoba et al. [29] in films highly populated with vortices. In this case, the reentrance mechanism is driven by vortex pinning and induced by a magnetic field. The vortices are confined in narrow constrictions by surface superconductivity, leading to a reentrant zero-resistance state.

Fermi surface reconstruction: High magnetic fields have induced the emergence of a superconducting state in the ferromagnet URhGe [30, 31]. This mechanism is explained by Yelland et al. [31] as a magnetic field-tuning of the Fermi surface, where Fermi surface pockets vanish at the zero-temperature field-induced Lifshitz transition (change of the Fermi surface without symmetry breaking).

Here, it is important to note that all these reentrance mechanisms are magnetic field dependent, with the exception of Josephson coupling.

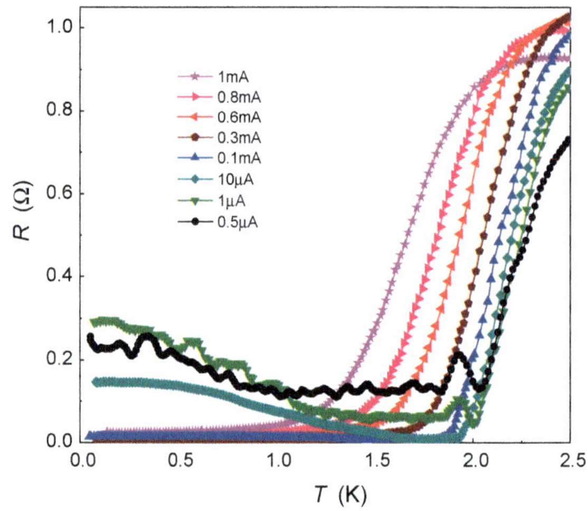

Fig. 7.5 Low temperature transport data $R(T)$ in $Na_{2-\delta}Mo_6Se_6$ crystal C for different applied currents $I = 0.5\,\mu A$ to $1\,mA$

7.3 Experimental Study of the Superconducting State

The following experimental data were obtained from a detailed study of crystal C in the dilution refrigerator and superconducting magnets. As we have seen in Chaps. 5 and 6, this crystal develops a stable $R \sim 0$ state at low temperature. The superconducting state is now investigated further to clarify the role of Josephson coupling on the superconducting phase coherence. As a side note, the color code used in the two previous chapters is not perpetuated here.

The superconducting transition in $R(T)$ for different applied currents is first presented in Fig. 7.5: notably, the transition and ground state strongly vary with increasing current. While it was previously suggested that crystal C had a coherent superconducting ground state, it seems to only be the case above a $\sim 100\,\mu A$ threshold current.

7.3.1 *Further Investigation of the 1D Superconducting Transition*

From band structure calculations and the analysis of the transition discussed in Sect. 6.3, the superconducting transition is known to be 1D at $T > T_J$, i.e. where no Josephson coupling is yet established [6, 36]. The analysis in the previous chapter suggests an important contribution to $R(T)$ from TAPS rather than QPS [36]. To identify a pure TAPS signature, transport data for four different applied currents are analyzed using Eq. 2.22 (Fig. 7.6).

In agreement with the previous observations, the TAPS model accurately reproduces $R(T)$ independent of the excitation current I: there is a universal onset of

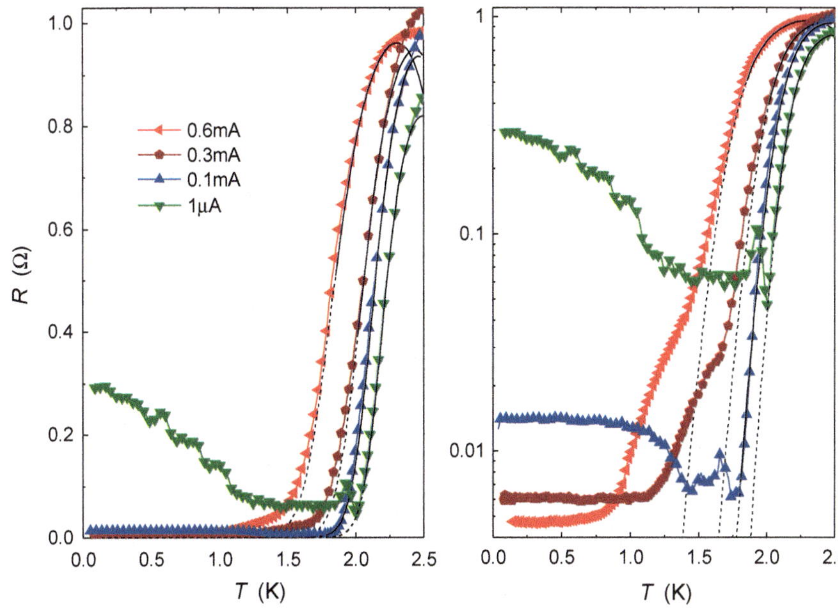

Fig. 7.6 Fits to TAPS model in transport measurements $R(T)$ for various applied currents $I = 1\,\mu$A to 0.6 mA, on linear (left) and semi-logarithmic (right) scales. The fits are shown as black lines switching to dotted lines where they deviate from the experimental data

fluctuating 1D superconductivity [6]. However, interesting features appear with the variation of I: the position of the hump is displaced to lower temperature as the current increases. For $I \leq 0.1$ mA, the resistance forms a plateau after the initial drop, before rising again at a lower temperature (reentrant superconductivity) [6]. In contrast, for $I > 0.1$ mA the resistance begins to diverge from the fitting at $T \sim 1.8$ K but continues to fall before saturating at low temperature (phase-coherent superconductivity) [6].

7.3.2 Observations of Reentrant Phase Coherence

The characterization of the BKT-like transition performed in Sect. 6.3.2 has demonstrated that long-range phase coherence is established in $Na_{2-\delta}Mo_6Se_6$ through a transition from fluctuating 1D superconductivity to a Josephson-coupled 3D superconducting ground state [6, 36].

A rise in resistance as $T \to 0$ for $I \leq 0.1$ mA (Fig. 7.5) suggests that the transverse phase coherence is fragile and reentrant. The evolution of $R(T)$ with current is detailed in Fig. 7.7 where an offset of 0.2 Ω between each curve highlights the reentrance.

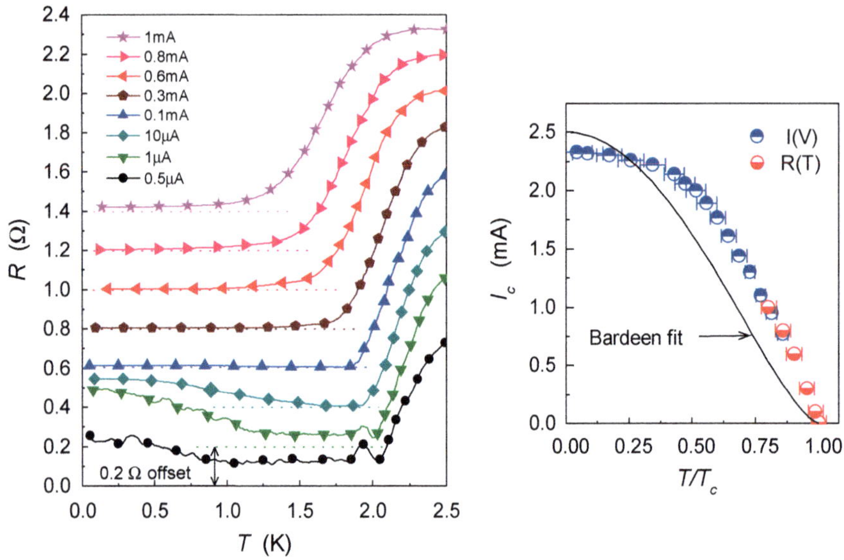

Fig. 7.7 Reentrant phase coherence in $R(T, I)$ and critical current $I_c(T)$. **Left**: $R(T)$ in zero magnetic field, acquired using 8 different currents $I = 0.5\,\mu$A to 1 mA. An offset of 0.2 Ω separates each curve for clarity; dashed lines indicate $R = 0$ for each data-set. **Right**: Temperature dependance of the critical current $I_c(T)$. Red circles correspond to data from $R(T)$ while blue data at larger I_c are extracted from Fig. 6.5a. The solid line shows the best $I_c(T)$ fit to Eq. 7.4

Three important trends may be identified in Fig. 7.7. First, the superconducting transition is conventionally suppressed to lower temperature as I increases. Secondly, for $T < 1$ K the resistance falls as I rises, i.e. phase coherence is established upon increasing I. Thirdly, the resistance rises (i.e. phase coherence is lost) upon reducing the temperature for $I \leq 0.1$ mA. Long range superconducting order is thus reentrant: it is only stable within a well-defined region of non-zero (T, I) phase space [6].

Figure 7.7 also shows $I_c(T)$ (right hand panel). $I_c(T)$ values are extracted from the $R(T)$ data (left hand panel) and the $V(I)$ data (Fig. 6.5), defined as $I_c(T) \equiv I(R(T) = 80\%\ R_N)$, where R_N is the normal-state resistance at T_{ons}.

$I_c(T)$ is fitted to the standard Bardeen relation derived for bulk superconductors [37]:

$$I_c(T) \propto \left(1 - \left(\frac{T}{T_c}\right)^2\right)^{3/2} \tag{7.4}$$

where $T_c = T_{ons}$. Interestingly, $I_c(T)$ does not follow Eq. 7.4 but remains unusually large at high temperature [6].

To determine whether phase coherence is also reentrant in magnetic field, the magneto-transport is measured in fields perpendicular and parallel to the c axis for two different currents: large current $I = 0.6$ mA (Fig. 7.8) and low current $I = 1\,\mu$A (Fig. 7.9). The measurements were performed in the dilution refrigerator combined

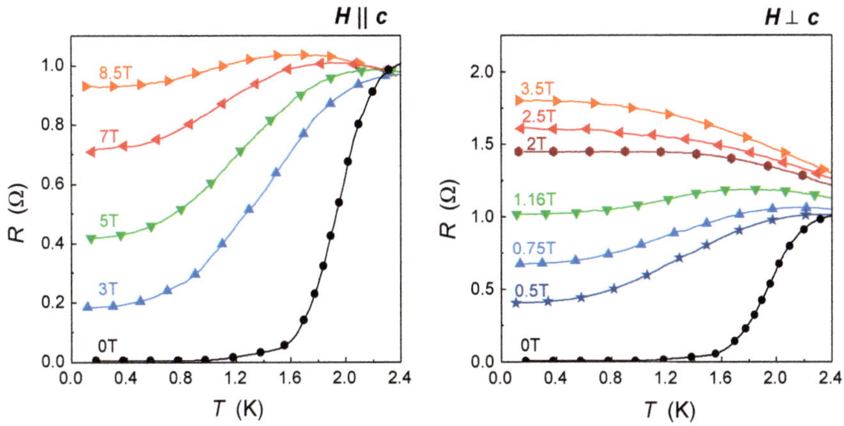

Fig. 7.8 Evolution of $R(T)$ in parallel ($H \parallel c$, **left**) and perpendicular ($H \perp c$, **right**) magnetic fields for large applied current $I = 0.6\,\text{mA}$. The values of the applied fields are indicated for each measurement

with the superconducting vector magnet where the achievable fields were $H_{\parallel} \leq 8.5\,\text{T}$ and $H_{\perp} \leq 3.5\,\text{T}$.

$R(T)$ for a large applied current ($I = 0.6$ mA, Fig. 7.8) are similar to those of a conventional superconductor and no reentrance is observed, although the transition is significantly broadened due to the large anisotropy and inhomogeneities. This signifies that transverse phase coherence has already been stabilized by the large applied current. Hence, the applied magnetic fields induce a purely destructive effect on superconductivity [6].

The anisotropy of the system is evident from the values of the applied fields: a peak is still visible at $H_{\parallel} = 8.5\,\text{T}$, while superconductivity is completely destroyed at $H_{\perp} = 2\,\text{T}$. The estimation of $H_{c2\perp,\parallel}(T)$ and superconducting coherence lengths are discussed in the analysis section.

On the other hand, $R(T)$ in low applied current ($I = 1\,\mu\text{A}$, Fig. 7.9) shows an unconventional behavior. In the presence of an applied magnetic field, $R(T)$ rises at low temperature, indicating a field-induced reentrance. The data in parallel magnetic fields show a shift of the reentrant minimum towards lower temperature with increasing field. In perpendicular field the data show quite complex behavior.

First, for $H_{\perp} = 0.75\,\text{T}$, $R(T)$ falls again below $T = 0.5$ K which corresponds to double-reentrance [13, 38]. Such behavior has been observed in granular systems and attributed to superconducting fluctuations, suppression of Cooper pair tunneling, and eventual macroscopic phase coherence as $R(T \to 0)$ [39, 40]. This cannot be the case in $Na_{2-\delta}Mo_6Se_6$ as both phase coherence and double-reentrance are absent as $T \to 0$ in the $H = 0\,\text{T}$ data. Instead, a possible explanation could come from mesoscopic fluctuations causing a divergent H_{c2} [41].

Secondly, an outstanding divergence of $R(T)$ for $T \leq 0.8$ K and $H_{\perp} \geq 2.5\,\text{T}$ is observed. This is attributed to magnetic field-induced Cooper pair localization, which has been predicted to occur in q1D materials when a field is applied perpendicular to

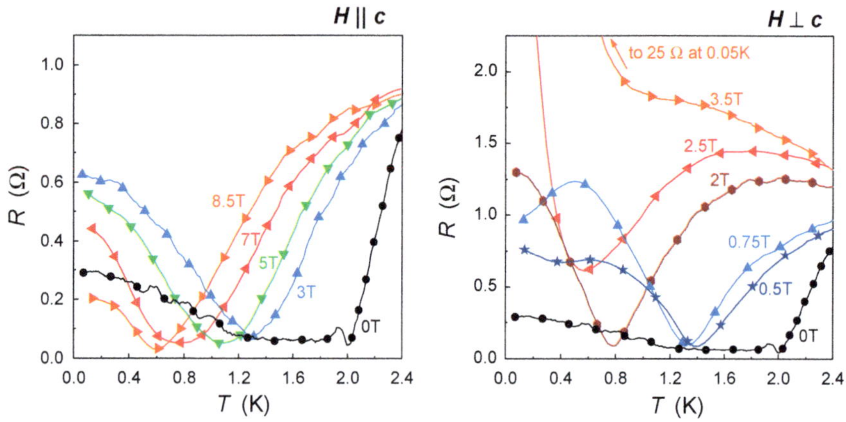

Fig. 7.9 Evolution of $R(T)$ in parallel ($H \parallel c$, **left**) and perpendicular ($H \perp c$, **right**) magnetic fields for low applied current $I = 1\,\mu$A. The values of the applied fields are indicated for each measurement

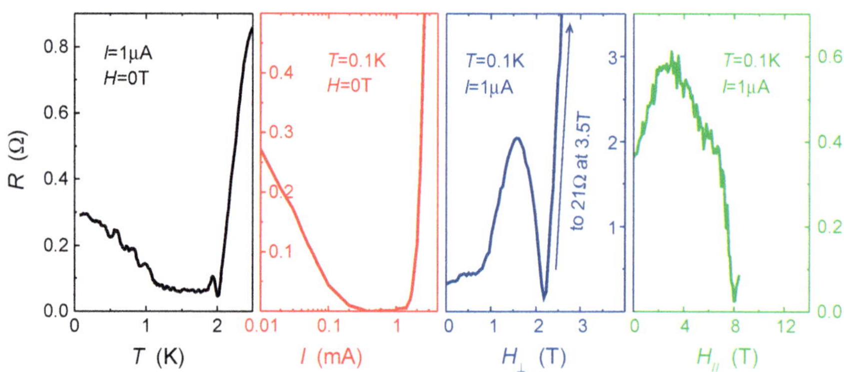

Fig. 7.10 Resistance $R(T, I, H_\perp, H_\parallel)$ in the superconducting phase. The respective fixed values of (T, H, I) are indicated in each dataset

the high-conductivity axis [42]. In the normal state, the conditions for localization are $\hbar\omega_c >> t_\perp$ and $k_B T << t_\perp$ (where ω_c is the cyclotron frequency $\mu_0 He/m_e$). Since electrons are paired for $T < T_{ons}$, single-particle hopping is replaced by Josephson tunneling as the transverse coupling mechanism: $t_\perp$ becomes $t_\perp^2/t_\parallel$ in the above condition. The estimation yields a minimum field of 1.3 T, in good agreement with the data [6]. This observation adds credence to the key role of Josephson coupling in establishing phase coherence.

The reentrance observed in $R(T)$ in Figs. 7.7 and 7.9 can be summarized by separately measuring $R(T)$, $R(H_{\perp,\parallel})$ and $R(I)$ (Fig. 7.10). In the superconducting phase of $Na_{2-\delta}Mo_6Se_6$, the resistance is minimized at non-zero (T, H, I): this indicates a reentrant behavior as a function of all three parameters [6].

The magnetoresistance in Fig. 7.10 is initially positive for $R(H_{\perp,\parallel})$, before falling steeply to the minimum value as Josephson coupling is established. If pair-breaking was taking place with applied field, transport would show a gradual, monotonic negative magnetoresistance prior to reentrance at low temperature [43]. Since no such feature is observed, quasiparticle tunneling is unlikely to play a major role in the reentrant behavior of $Na_{2-\delta}Mo_6Se_6$ [6].

7.4 Analysis

7.4.1 Determination of H_{c2} and $\xi(0)$

As expected for a q1D superconductor, the magneto-transport in Figs. 7.8 and 7.9 varies strongly with the field orientation, regardless of the applied current. The anisotropy is quantified via the temperature dependence of the upper critical fields $H_{c2\parallel,\perp}(T)$ and plotted in Fig. 7.11.

The critical fields are defined as $H_{c2}(T) \equiv H(R(T) = 80\% R_N)$ to consistently characterize the entire superconducting phase, since the error in $T_{ons}(H, I)$ is large for high (H, I) and $T_{BKT}(H, I)$ cannot accurately be extracted across all (H, I) datasets.

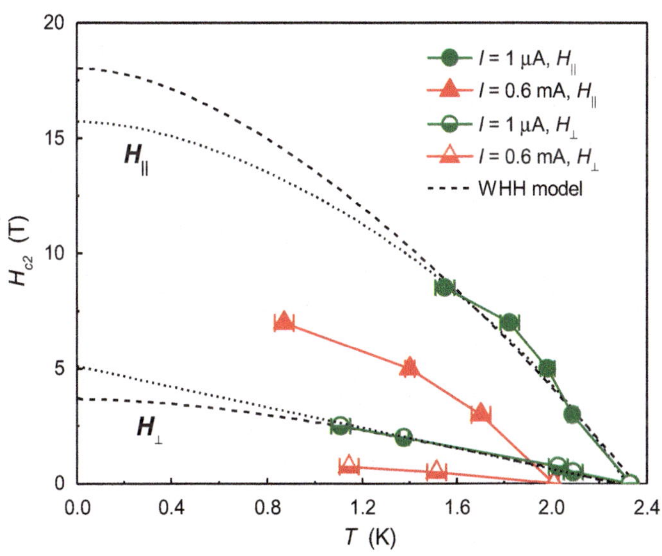

Fig. 7.11 Determination of H_{c2} from the $R(T, H)$ data defined as $H_{c2}(T) \equiv H(R(T) = 80\% R_N)$. The dotted/dashed lines are fits representing upper/lower limits to H_{c2}. The WHH fits were performed by A. P. Petrović

Table 7.1 Experimental values of H_{c2} and $\xi(0)$ parallel ($\|$) and perpendicular ($\perp$) to the chains in $Na_{2-\delta}Mo_6Se_6$

	$\|$		$\perp$	
	Min	Max	Min	Max
H_{c2} (T)	15.7	18.0	3.66	5.02
$\xi(0)$ (nm)	14.4	21.0	4.28	4.58

To estimate the values of $H_{c2\|}(0)$ and $H_{c2\perp}(0)$, the Werthamer–Helfand–Hohenberg (WHH) model [35] is used to fit the low current ($I = 1\ \mu A$) data:

$$\ln\frac{1}{t} = \sum_{n=-\infty}^{\infty} \frac{1}{|2n+1|} - \left[|2n+1| + \frac{\overline{h}}{t} + \frac{(\alpha\overline{h}/t)^2}{|2n+1| + (\overline{h}+\lambda_{so})/t}\right]^{-1} \tag{7.5}$$

Where $t = T/T_c$, $\alpha = -5.2758 \times 10^{-5} \left.\frac{dH_{c2}}{dT}\right|_{T_c}$ is the Maki parameter, $\overline{h} = -(4/\pi^2)H_{c2}/\left.\frac{dH_{c2}}{dt}\right|_{t=1}$ the reduced magnetic field and λ_{so} the spin-orbit coupling.

The errors in the fits are defined as follows: for $H_{c2\|}(T)$, the WHH equations are calculated with $H_{c2}(T + \delta T)$, where δT is the error on the temperature axis; for $H_{c2\perp}(T)$, the lower limit is defined by the WHH fit while the upper limit assumes that $H_{c2}(T)$ is linear at all temperatures as was suggested in $Tl_{2-\delta}Mo_6Se_6$ [44]. The results of the fits can be read on the $T = 0$ axis in Fig. 7.11 and are summarized in Table 7.1.

Values of the coherence lengths $\xi_{\|}(0)$ and $\xi_{\perp}(0)$ can be extracted using the estimates of $H_{c2\|,\perp}(T = 0)$ and GL theory (Eq. 2.17). The values obtained are presented in Table 7.1 and yield an anisotropy $\epsilon \equiv \xi_{\|}(0)/\xi_{\perp}(0)$ of $3.14 \le \epsilon \le 4.90$ [6].

The value of the estimated ϵ is lower than the 12.6 reported for $Tl_{2-\delta}Mo_6Se_6$ and 17 for $In_{2-\delta}Mo_6Se_6$ [45], although $t_{\perp}$ is smaller in $Na_{2-\delta}Mo_6Se_6$. The variation in the anisotropies can be explained as follows [6]:

First, disorder from the high Na vacancy density strongly suppresses $\xi_{\|}(0)$ and hence ϵ. This is apparent by the orbitally-limited values for $H_{c2\perp}$ in $Tl_{2-\delta}Mo_6Se_6$ and $In_{2-\delta}Mo_6Se_6$: 0.47 and 0.25 T respectively [45], an order of magnitude lower than the 3.7–5 T measured for $Na_{2-\delta}Mo_6Se_6$.

Second, the estimated $H_{c2\|} \approx$ 16–18 T exceeds the weak-coupling BCS Pauli limit $H_P \equiv \frac{\Delta_0}{\sqrt{2}\mu_B} = 5.0$ T by a factor > 3. This means that paramagnetic rather than orbital limiting must be suppressing superconductivity for $H \| c$. Therefore, GL theory may only provide an upper limit for $\xi_{\perp}(0)$.

7.4.2 Estimation of the Filamentary Diameter

The results of the TAPS+QPS fitting procedure performed in Sect. 6.3 allow us to estimate the typical diameter d_F of the superconducting filaments in $Na_{2-\delta}Mo_6Se_6$

crystal C [6]. The effective sample length was defined $L_{eff} = Lm/n$, where L is the length of one unit in a $m \times n$ array of filaments. First, let us assume that current flows homogeneously through the crystal. The size of the sample is set as $Lm = w$ (where $w = 0.4$ mm is the voltage contact separation in crystal C) and the TAPS fitting parameter $L_{eff}/\xi(0)$ is equal to $Lm/\left(n\xi_{\parallel}(0)\right)$. Using the experimental $\xi_{\parallel}(0) = 14 \rightarrow 21$ nm (Table 7.1), the number of filaments in a typical crystal cross-section n can be deduced and hence the filament diameter can be estimated. For $I = 1\,\mu\text{A}$, $n = 3.5 \rightarrow 2.3 \times 10^6$ is obtained and using the crystal cross-section $A \sim 8 \times 10^{-9}\,\text{m}^2$, a maximum filament diameter $d_F = 54 \rightarrow 66$ nm is deduced [6].

However, it has been demonstrated in Sect. 4.4 that the current is largely anisotropic in a disordered q1D material (Fig. 4.12). The typical current-carrying fraction of the crystal cross-section is therefore small ($\ll A$), leading to a considerable reduction in d_F [6].

To estimate the reduction in d_F, the experimentally-determined $T_{BKT} = 1.71$ K is used to extract a Josephson energy [46] $E_J \equiv 2k_B T_{BKT}/\pi = 94\,\mu\text{eV}$ and the inter-filamentary critical current $I_c^f \equiv 2eE_J/\hbar = 46$ nA. Defining a total critical current $I_c(T = 1.71\,\text{K}) = 1.2\,\text{mA}$ from the $V(I)$ data presented in the previous chapter (Fig. 6.5), the number of current-carrying filaments in the crystal cross-section is estimated ($n_J \equiv I_c/I_c^f$) 2.6×10^4, which is considerably smaller than the $n = 3.5 \rightarrow 2.3 \times 10^6$ filaments deduced from the TAPS fitting. The ratio of these values n_J/n provides an approximate fraction of the total crystal cross-section carrying a supercurrent, implying a typical filamentary diameter $d_F = 0.41 \rightarrow 0.74$ nm. The Se-Se diameter of a single $(Mo_6Se_6)_\infty$ chain is 0.60 nm and the hexagonal lattice parameter $a = 0.86$ nm: for $T_{BKT} < T < T_{ons}$, single $(Mo_6Se_6)_\infty$ chains are behaving as 1D superconducting filaments, as expected from the electronic band structure [6].

7.4.3 *Mechanism for Reentrant Superconductivity in $Na_{2-\delta}Mo_6Se_6$*

From the experimental data, the mechanism controlling the reentrant phase coherence in $Na_{2-\delta}Mo_6Se_6$ can easily be identified. The observation of reentrance at $H = 0$ and $I << I_c$ (Figs. 7.5, 7.6, 7.7, 7.8, 7.9 and 7.10) directly absolves mechanisms such as exchange-field compensation (Jaccarino–Peter effect), FFLO state formation, vortex pinning, Fermi surface reconstruction and magnetic impurities from being responsible for reentrance. In addition, the effect of magnetic impurities is unlikely to play a major role in $Na_{2-\delta}Mo_6Se_6$, since the crystals show no evidence for any significant magnetic impurity content (Fig. 5.13). The only known mechanism left to explain reentrant superconductivity in $Na_{2-\delta}Mo_6Se_6$ is by Josephson coupling.

The experimental data in Sect. 6.3 and the band structure calculations (Sect. 4.4) have also shown that the dimensional crossover in the superconducting transition is due to Josephson coupling between the (Mo_6Se_6) chains [6, 36]. In conclusion, the reentrant phase coherence in $Na_{2-\delta}Mo_6Se_6$ takes place by Josephson coupling [6], as previously observed in inhomogeneous materials.

7.4.4 Calculations of the Josephson Energy

The origin of the reentrance can be clarified by estimating the (T, H, I) dependence of E_J within the analytical framework proposed by Belevtsev et al. for inhomogeneous superconductors (Eqs. 7.1, 7.2, 7.3). In the filamentary $Na_{2-\delta}Mo_6Se_6$ [6]:

$$E_J = \frac{\pi\hbar}{4e^2 R_t}\Delta(T)\tanh\left(\frac{\Delta(T)}{2kT}\right) \quad (7.6)$$

where $R_t = E_a/e^2\omega_T$ is taken as the tunneling resistance between two superconducting filaments, with the characteristic tunneling frequency:

$$\omega_t \sim A\exp\left[-\left(\frac{2m_e s^2\varphi_0}{\hbar^2}\right)^{1/2}\right]\cdot\exp\left[-\frac{E_a}{k_B T}\right] \quad (7.7)$$

where $\varphi_0 = (\Phi - eV)$ is the effective barrier height.

In granular superconductors for which these equations were initially derived, E_a is the capacitive charging energy of each grain. In $Na_{2-\delta}Mo_6Se_6$, there is no charging energy of the filaments: the filaments are conductive and the conduction is strongly influenced by Na vacancy disorder [6, 36]. The normal state of this material is strongly localized at low temperature (Sect. 5.4). The system can be understood as 1D superconducting filaments in which the normal state is governed by strongly localized electrons displaying VRH transport. In the model for E_J, the activation energy E_a can therefore be characterized using VRH theory. The VRH model is extrapolated to describe the normal state magnetoresistance [6, 47]:

$$R(H) \propto \exp\left[\frac{T_0(H)}{T}\right]^{\frac{1}{1+d}} \quad (7.8)$$

Here, the normal state $R(T)$ and $R(H)$ (Fig. 7.12) can be used to extract the VRH fitting parameters. The fit to $R(T)$ is identical as the one discussed in Chap. 5 (Fig. 5.7, crystal C) and the fit to $\ln R(H)$ is performed using Eq. 7.8.

It is now possible to express the VRH activation energy E_a as:

$$E_a(T, H) \equiv \frac{\partial \ln R}{\partial (k_B T)^{-1}} = \frac{k_b T_0(H)}{1+d}\left(\frac{T_0(H)}{T}\right)^{\frac{1}{1+d}-1} \quad (7.9)$$

In summary, electron localization induced by disorder controls $E_a(T, H)$, and therefore R_t [6]. This is presented in Fig. 7.13: in an inhomogeneous q1D superconductor composed of finite-length dirty nanofilaments, the conduction band electrons become localized at low temperature. The localized wavefunctions are depicted in the inset, limited by an exponential decay envelope. Their overlap along the a-axis (green shading) is small when compared to that along the filaments, and the transport is mostly

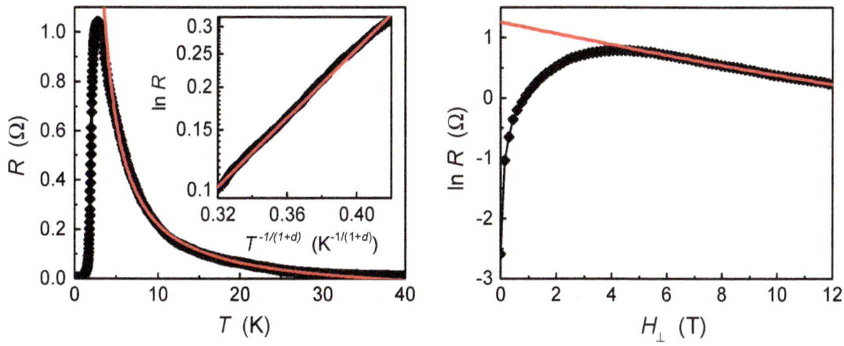

Fig. 7.12 VRH fits (black lines) to the normal-state transport in $Na_{2-\delta}Mo_6Se_6$. **Left**: Zero-field $R(T)$ fitted with VRH. **Right**: $\ln R(H_\perp)$ at $T = 1.8$K. The high field strong negative magnetoresistance is fitted to Eq. 7.8

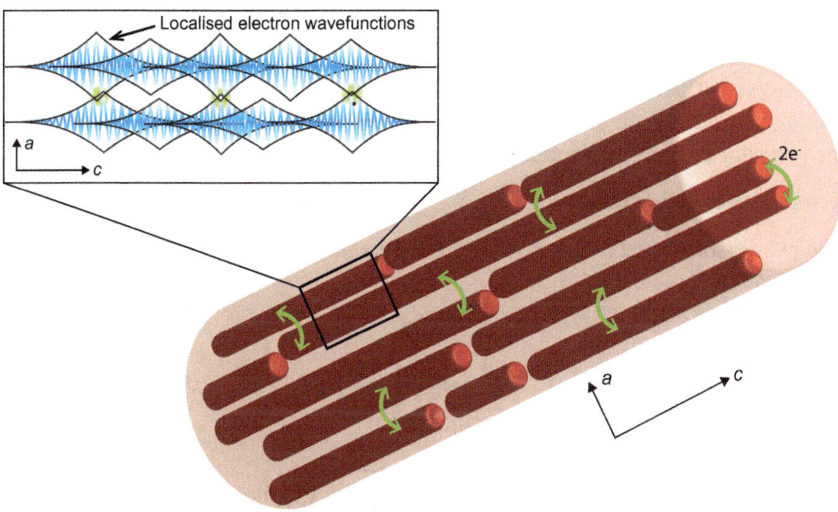

Fig. 7.13 Cartoon illustrating the role of electron localization in the E_J analysis. The 1D filaments (red) become phase-coherent when Josephson coupling (green arrows) is established. **Inset**: The overlap (green shading) of localized electron wavefunctions initiates coupling between the filaments

1D. At non-zero temperature, field or current, the electrons become delocalized and Josephson tunneling (green arrows) stabilizes phase coherence [6].

Reentrant superconductivity in $Na_{2-\delta}Mo_6Se_6$ can now be analyzed using the expression of E_J and $E_a(T, H)$. A reentrant superconducting behavior corresponds to a stabilized Josephson coupling between the 1D filaments, hence characterized by a maximum in E_J [6]. A peak in E_J can therefore be tracked independently as a function of T, H and I.

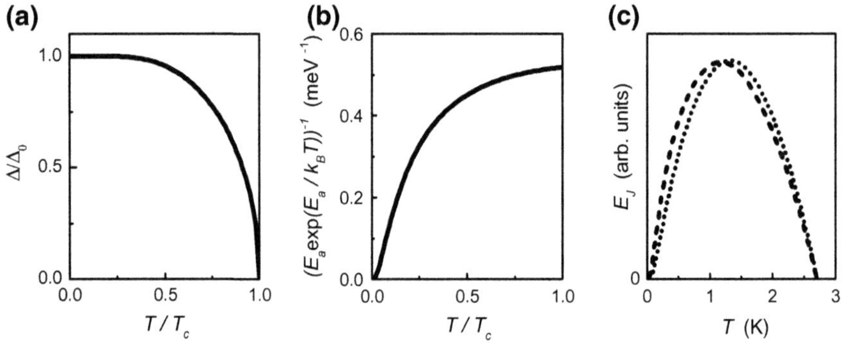

Fig. 7.14 Factors controlling $E_J(T)$. **a** Renormalized temperature-dependent BCS gap Δ/Δ_0. **b** $(E_a(T)\exp(E_a(T)/k_BT))^{-1}$ which is proportional to $R_t^{-1}(T)$. **c** $E_J(T)$ in arbitrary units, where the upper and lower limits (dashed/dotted lines, respectively) are obtained from the VRH-extracted T_0 and d parameters

However, it is not possible to extract numerical values of $E_J(T, H, I)$: several scaling parameters are unknown [6]. For this reason, only a qualitative analysis of the behavior of E_J as a function of each parameter is performed.

Temperature Dependence

The expression for $\Delta(T)$ is retrieved from a numerical solution of the s-wave BCS gap equation:

$$\Delta_k(T) = -\frac{1}{2}\sum_{k'} V_{kk'}^{(0)} \frac{\Delta_{k'}}{\sqrt{|\varepsilon_{k'}^2| + |\Delta_{k'}^2|}} \tanh \frac{\sqrt{|\varepsilon_{k'}^2| + |\Delta_{k'}^2|}}{2T} \tag{7.10}$$

where $V_{kk'}$ is the pairing potential and $\sqrt{|\varepsilon_{k'}^2| + |\Delta_{k'}^2|}$ the quasiparticle excitation energy (Eq. 2.7). From BCS theory (Chap. 2, Fig. 2.3), we know that $\Delta(T)$ decreases with increasing temperature, yielding $\Delta(T) = 0$ at $T = T_c$ (Fig. 7.14a).

$R_t^{-1}(T) \propto (E_a(T)\exp(E_a(T)/k_BT))^{-1}$ is estimated using the expression of $E_a(T, H)$ in Eq. 7.9 and the experimentally-determined VRH parameters T_0 and d. Due to electron localization, $R_t^{-1}(T)$ increases with increasing temperature (Fig. 7.14b).

The interplay between $\Delta(T)$ and R_t^{-1} forms a peak in $E_J(T)$ (Fig. 7.14c) and indicates a reentrant superconducting state at $T \neq 0$ [6]. Errors in the estimated $E_J(T)$ are extracted from the errors in T_0 and d of the VRH fit and are shown as the lower (dotted line) and upper (dashed line) limits in the location of the peak.

Field Dependence

$\Delta(H)$ is approximated using the following expression [48]:

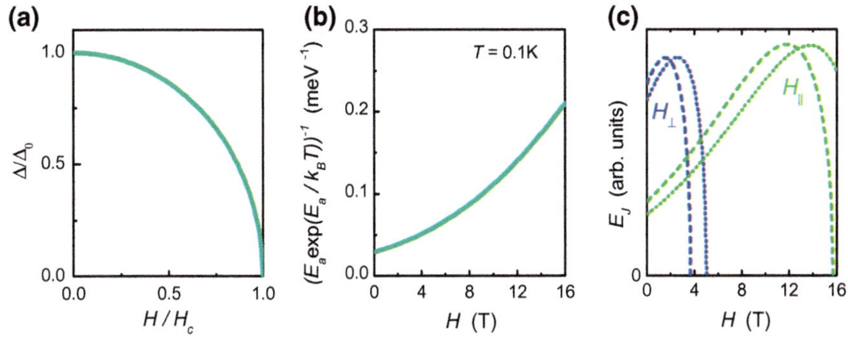

Fig. 7.15 Factors controlling $E_J(H)$. **a** Renormalized field-dependent gap Δ/Δ_0 extracted from [48]. **b** $(E_a(H)\exp(E_a(H)/k_BT))^{-1}$, which is proportional $R_t^{-1}(H)$. **c** $E_J(H)$ in arbitrary units, where the upper and lower limits (dashed/dotted lines, respectively) are obtained from the WHH fits for $H_{c2\parallel,\perp}$

$$\Delta(H) = \Delta_0 \left(1 - \left(\frac{H}{H_{c2}}\right)^2\right)^{\frac{1}{2}} \tag{7.11}$$

where H_{c2} is determined from WHH fits (Fig. 7.11). The result is plotted in Fig. 7.15a.

$R_t^{-1}(H) \propto (E_a(H)\exp(E_a(H)/k_BT))^{-1}$ is estimated using the expression of $E_a(T, H)$ in Eq. 7.9. The experimentally observed negative magnetoresistance (right hand panel in Fig. 7.12) can be modeled using Eq. 7.8. Once the superconductivity and fluctuation paraconductivity have been suppressed ($H > 5.2$ T), $R(H)$ can be fitted using $a + b\exp(-cH)$. This leads to:

$$a + b\exp(-cH) = R_0 \exp\left(\frac{T_0(H)}{T}\right)^{\frac{1}{1+d}} \tag{7.12}$$

which can be used to obtain $T_0(H)$. Substituting the numerical expression for $T_0(H)$ into Eq. 7.9, a field-dependent activation energy $E_a(H)$ is obtained. Similar to the temperature dependence, $R_t^{-1}(H)$ increases due to electron delocalization with increasing field (Fig. 7.15b).

The interplay between $\Delta(H)$ and R_t^{-1} forms a peak in $E_J(H)$ (Fig. 7.15c), indicating a reentrant superconducting state at $H \neq 0$ [6]. Here, the upper and lower WHH limits for $H_{c2\parallel,\perp}$ provide the upper and lower limits (dashed/dotted lines) in $E_J(H_{\parallel,\perp})$.

Current Dependence

The normal state resistance in $Na_{2-\delta}Mo_6Se_6$ is ohmic: $V = IR_N$ where $R_N \approx 1\ \Omega$, as observed in the $V(I)$ curve (Fig. 6.5). Therefore, one can approximate $I \sim V$ and substitute I for V in this analysis.

The current dependence of the superconducting gap $\Delta(I)$ is still an open question theoretically. However, it is possible to derive an approximate relation using GL theory [49]:

$$\frac{|\Psi|^2}{\Psi_\infty^2} = 1 - \left(\frac{\xi m_e v_s}{\hbar}\right)^2 \tag{7.13}$$

where Ψ is the GL order parameter, v_s the superfluid velocity, and $|\Psi|^2 = n_s$ (Sect. 2.1.4). Since $j_s = 2en_s v_s$, j_s and v_s have a roughly linear relationship (except close to J_c).

Next, we approximate:

$$\frac{|\Psi|^2}{\Psi_\infty^2} \approx \left(\frac{\Delta(I)}{\Delta_0}\right)^2 \tag{7.14}$$

The gap closes for $I \equiv I_c$ and so $\Delta(I)$ can be estimated using:

$$\Delta(I) = \Delta_0 \left(1 - \left(\frac{I}{I_c}\right)^2\right)^{\frac{1}{2}} \tag{7.15}$$

This relation is similar to Eq. 7.11 for $\Delta(H)$. The result is plotted in Fig. 7.16a.

The current dependence of the tunneling resistance is obtained by estimating $R_t^{-1} \propto \omega_t \propto \exp\left[-\left(\frac{2m_e s^2 \varphi_0}{\hbar^2}\right)^{1/2}\right]$. The barrier height $\varphi_0 = \Phi - eV$ and thickness s are both unknown and require heavy approximations [6].

In a BKT transition, the Josephson energy can be expressed as $E_J = \frac{2}{\pi} k_B T_{BKT}$ at T_{BKT}. $\Delta(T = T_{BKT})$ and $E_a(T = T_{BKT})$ are evaluated for $Na_{2-\delta}Mo_6Se_6$ using the VRH parameters T_0 and d. The obtained numerical values are substituted into the expression for E_J (Eq. 7.6) setting $A = 1$, which yields $\frac{2m_e s^2 \varphi_0}{\hbar^2} = 775$.

The barrier thickness s has to be determined: it should lie above the interchain separation ~ 0.64 nm and below the 1D localization length. This is approximated [50] as $\xi_L \equiv 4/k_B T_0 N(E_F)^{1D} = 380$ nm (where $N(E_F)^{1D} = 1.07 \times 10^{28}\ \mathrm{J^{-1}m^{-1}}$ is the 1D density of states, estimated using the value for $In_{2-\delta}Mo_6Se_6$ [45]). A realistic approximation of s is chosen as the logarithmic average of these values: 16 nm. Using this value, the effective barrier height is $\varphi_0 = \Phi - eV = 0.1155$ eV, which is physically reasonable. To complete the approximation, the value of $I = 6 \times 10^{-4}$ A is selected, which corresponds to the applied current in the measurement where T_{BKT} was extracted. As $V \sim I$, this sets $\Phi = 0.1161$ eV.

The behavior of $\exp\left[-\left(\frac{2m_e s^2 \varphi_0}{\hbar^2}\right)^{1/2}\right]$ is plotted in Fig. 7.16b.

The interplay between $\Delta(I)$ and R_t^{-1} results in a peak in $E_J(I)$ at $I \neq 0$ (Fig. 7.16c) [6]. In this case, no upper and lower limits are presented for $E_J(I)$ due to the large degree of approximation. The main aim here is to demonstrate that a peak in $E_J(I)$ develops for non-zero I.

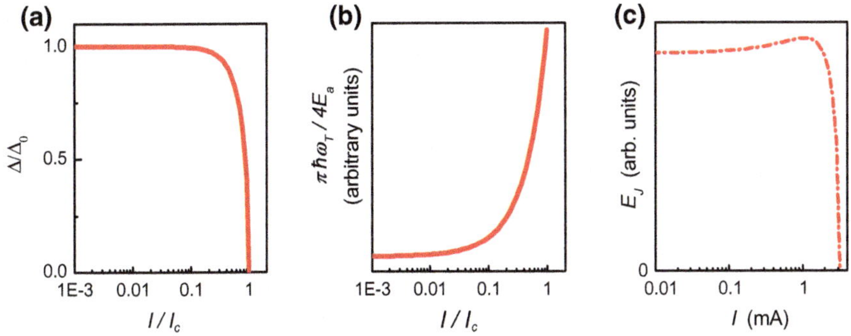

Fig. 7.16 Factors controlling $E_J(I)$. **a** Renormalized current-dependent gap Δ/Δ_0 using GL theory. **b** $\pi\hbar\omega_T(I)/4E_a$, which is proportional to $R_t^{-1}(I)$. **c** $E_J(I)$ in arbitrary units, where no error was calculated due to the highly approximate calculation procedure

7.4.5 Correspondence Between Experimental Data and Calculations

The result of these separate calculations is presented in Fig. 7.17, where the Josephson energy is plotted independently as a function of temperature, current and both field orientations. E_J is plotted in arbitrary units as several scaling parameters are unknown. Nevertheless, the approximate calculations provide qualitative information for $E_J(T, I, H_\perp, H_\parallel)$: for each parameter independently, a peak develops in E_J due to the interplay between the superconducting gap and the tunneling resistance between superconducting filaments [6]. The experimental data for $R(T, I, H_\perp, H_\parallel)$ are plotted below, similar to the data in Fig. 7.10.

The correspondence between the peak in E_J and the minimum in the resistance is highlighted by the vertical dashed lines. This indicates that the reentrant behavior in $Na_{2-\delta}Mo_6Se_6$ is controlled by Josephson coupling between superconducting filaments [6]. An exception occurs however for $H \parallel c$, where the calculated E_J peak takes place at a higher field. The H_{c2} values were calculated using the WHH fits (Fig. 7.11), which consider orbital limiting. However, as we have seen, it is more likely that paramagnetic limiting controls H_{c2} in $Na_{2-\delta}Mo_6Se_6$: the values used in the calculation are most certainly overestimated [6].

The reentrant experimental data presented in this chapter are summarized by the extent of phase coherence in $Na_{2-\delta}Mo_6Se_6$ as mapped in a (T, H, I)-phase diagram (Fig. 7.18). The values of $H_{c2}(T)$ and $I_c(T)$ plotted in Figs. 7.7 and 7.11 are represented by circles: they accurately trace the evolution of the superconducting transition.

To describe the loss of phase coherence at low (T, H, I), a reentrance threshold temperature $T_R(H, I)$ is defined as the minima in $R(T, H)$ in Figs. 7.7, 7.9, and 7.10, represented with stars. At temperatures below $T_R(H, I)$ the filaments are superconducting but phase-incoherent and $Na_{2-\delta}Mo_6Se_6$ exhibits a finite resistance,

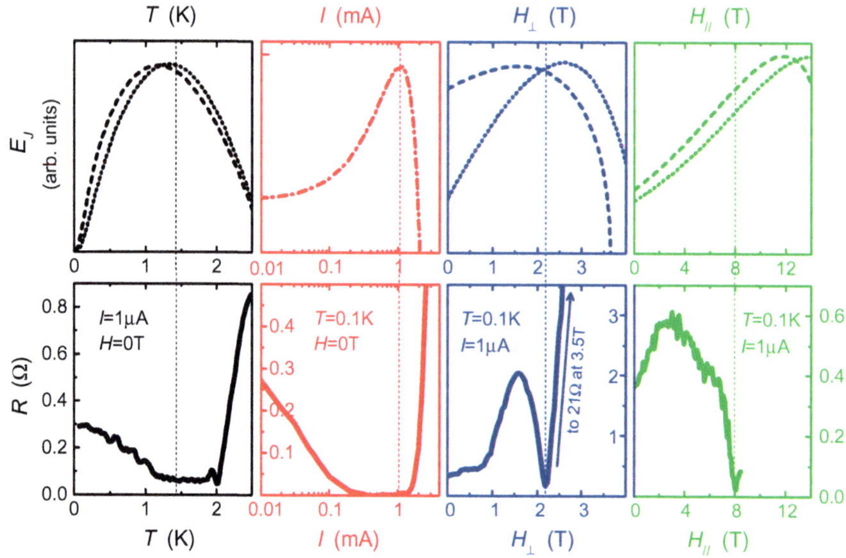

Fig. 7.17 **Top**: $(T, I, H_{\perp,\parallel})$-dependence of the estimated Josephson energy E_J. **Bottom**: $(T, I, H_{\perp}, H_{\perp,\parallel})$-dependence of the experimental resistance R, similar to Fig. 7.10. A vertical dashed line highlights the correspondence between E_J and R for each parameter

characteristic of fluctuating 1D superconductivity. In this phase diagram, the volume defined by $H_{c2}(T)$, $I_c(T)$, and $T_R(H, I)$ indicates a reentrant shell of phase-coherent superconductivity.

The E_J calculations are summarized in Fig. 7.19. Here, the theoretical $\Delta(T, H, I)/\Delta_0$ is first plotted in 3D shading. Calculated values of $T_R(H, I)$ are obtained by tracking the location of the peaks in E_J within the (H, T) and (I, T) planes (yellow and white data points, respectively). A similar trend is observed between the experimental and calculated phase diagrams.

However, the $(0, H, I)$-plane of the calculated phase diagram is colored in orange to indicate the limitation of the E_J model for $Na_{2-\delta}Mo_6Se_6$ [6]. The normal-state VRH activation energy is used as $E_a(T, H)$ in the expression of E_J, which implies that R_t would always diverge to infinity as $T \to 0$ (from the $\exp(-E_a/k_BT)$ term in ω_t, Eq. 7.7). In contrast, the experimental data indicate stable reentrant phase coherence at large fields or currents, even at low temperatures: this is incompatible with a divergent R_t.

To satisfy what is observed in the experimental data, the zero-temperature normal state of $Na_{2-\delta}Mo_6Se_6$ must be metallic, not insulating. This is supported by the position of crystal C on the extended-side of the observed mobility edge (Sect. 5.4) [36]. In this crystal, the electron wavefunctions are in an extended state (although very close to the mobility edge) and the VRH equation merely provides an approximate fit over a finite non-zero temperature range [6, 36]. Hence, $R(T \to 0)$ does not diverge to infinity but remains metallic.

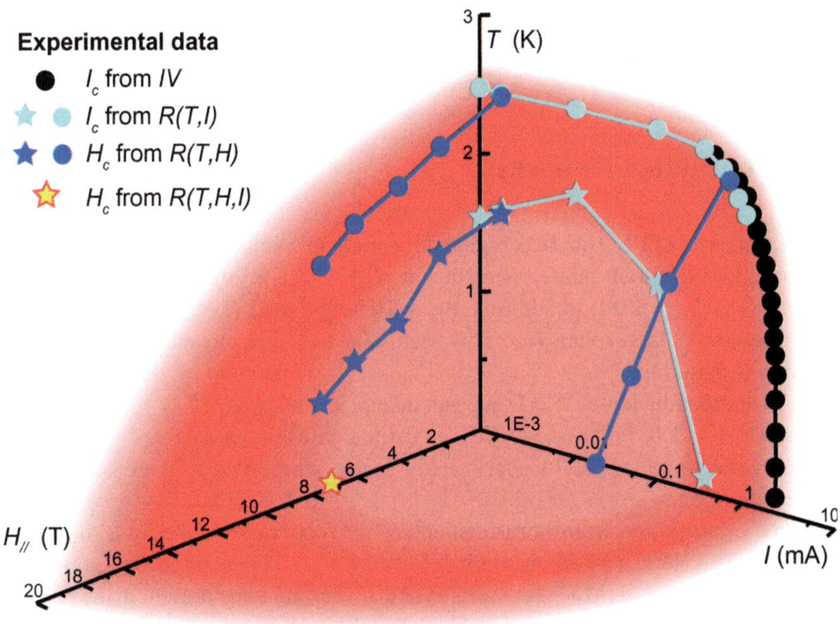

Fig. 7.18 Experimental phase diagram illustrating H_{c2}, I_c (circles) and $T_R(H, I)$ (stars). The dark red shading is a guide to the eye, highlighting the shell of reentrant phase coherence at non-zero (T, H, I). Replacing $H_\parallel$ by $H_\perp$ leads to a similar diagram with the H axis renormalized by the anisotropy ϵ

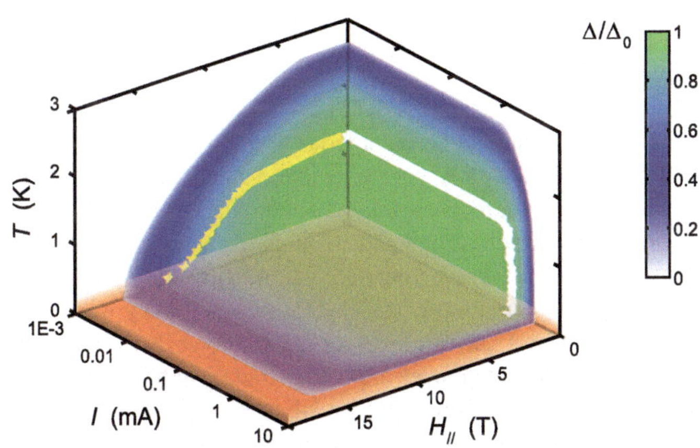

Fig. 7.19 Calculated phase diagram showing $T_R(H, I)$ (yellow/white data points) and the normalized pairing energy Δ/Δ_0. The orange-colored plane indicates a failure in the E_J model when $T \to 0$. Calculations were performed by A. P. Petrović

Nevertheless, the physical concept which underlies the model (i.e. the formation of peaks in E_J at non-zero (T, H, I)) remains valid [6].

7.5 Concluding Remarks

The presence of a FFLO phase at high fields cannot be excluded in $Na_{2-\delta}Mo_6Se_6$ [6]. The observed reentrant phase coherence at $H_{\|} = 8.5\,T$ is considerably higher than $H_P \equiv 1.84\,T_{ons} = 5.0\,T$. In addition, the WHH-estimated $H_{c2//} \sim 16$–$18\,T$ is more than three times larger than H_P. This seems far too high to only be explained by spin-orbit scattering.

In a superconductor, a FFLO state can take place below $0.55T_{ons}$ [51] if the Maki parameter $\alpha > 1.6$. In $Na_{2-\delta}Mo_6Se_6$, an estimated Maki parameter $4.4 \leq \alpha \leq 5.1$ encourages a possible FFLO scenario at low temperature [6]. This therefore deserves further attention.

Some contribution from intra-filamentary defects could also play a role in the reentrance [6]. Here, intra-filamentary defects refer to Josephson coupling across barriers within individual $(Mo_6Se_6)_\infty$ chains and are certain to be present in the crystals. These defects are probably responsible for the plateaus observed in $R(T)$ at low current (Fig. 7.7) [6]. However, supercurrents can percolate around such barriers without large resistive losses, provided that the chains are phase-coherent. Since phase coherence is established at $T_{BKT} > T_R$, the rise in the resistance below $T_R(H, I)$ must correspond to the loss of transverse phase coherence [6].

Finally, T_{ons} in $Na_{2-\delta}Mo_6Se_6$ is low due to the combination of a small DoS at the Fermi level and weak e-ph coupling. However, the electron delocalization mechanism responsible for the reentrance remains active at temperatures at least an order of magnitude higher than T_{ons} (Fig. 7.12). This means that in the presence of a pairing interaction, there should be no obvious thermal limitation to the reentrance mechanism. It has been shown that negative magnetoresistance and $dR/dT < 0$ can persist up to room temperature in disordered nanomaterials [52].

In summary, reentrant phase coherence in $Na_{2-\delta}Mo_6Se_6$ was observed as a function of temperature, applied magnetic field and current. This was successfully described by Josephson coupling, where the Josephson energy was calculated as a function of each parameter. The results of the calculations closely match the experimental data. Together with recent work indicating giant T_c enhancements in superconducting nanoparticles [53–56], this inbuilt resilience to phase fluctuations supports the assembly of dirty nanowire arrays as an attractive route towards synthesizing new functional superconductors.

References

1. V.F. Gantmakher, V.T. Dolgopolov, Superconductor - insulator quantum phase transition. Phys. Uspekhi **53**(1), 1–49 (2010)
2. I. Beloborodov, A. Lopatin, V. Vinokur, K. Efetov, Granular electronic systems. Rev. Mod. Phys. **79**(2), 469–518 (2007)
3. R. Kleiner, F. Steinmeyer, G. Kunkel, P. Müller, Intrinsic Josephson effects in Bi2Sr2CaCu2O8 single crystals. Phys. Rev. Lett. **68**(15), 2394 (1992)
4. P.J.W. Moll, X. Zhu, P. Cheng, H.-H. Wen, B. Batlogg, Intrinsic Josephson junctions in the iron-based multi-band superconductor (V2Sr4O6)Fe2As2. Nat. Phys. **10**(9), 644–647 (2014)
5. H.J. Schulz, Long-range Coulomb interactions in quasi-one-dimensional conductors. J. Phys. C Solid State Phys. **16**, 6769–6787 (1983)
6. D. Ansermet, A.P. Petrović, S. He, D. Chernyshov, M. Hoesch, D. Salloum, P. Gougeon, M. Potel, L. Boeri, O.K. Andersen, C. Panagopoulos, Reentrant phase coherence in superconducting nanowire composites. ACS Nano **10**(1), 515–523 (2016)
7. T. Lin, X. Shao, M. Wu, P. Hor, X. Jin, C. Chu, N. Evans, R. Bayuzick, Observation of a reentrant superconducting resistive transition in granular BaPb0.75Bi0.25O3 superconductor. Phys. Rev. B **29**(3), 1493–1496 (1984)
8. E. Simanek, Effect of charging energy on transition temperature in granular superconductors. Solid State Commun. **31**, 419–421 (1979)
9. K.B. Efetov, Phase transitions in granulated superconductors. Sov. Phys. JETP **51**(5), 1015–1022 (1980)
10. E. Simanek, Reentrant phase diagram for granular superconductors. Phys. Rev. B **32**(1), 500–502 (1985)
11. B.I. Belevtsev, Y.F. Komnik, A.V. Fomin, Reentrant phenomena in superconductivity of ultrathin amorphous bismuth films. J. Low Temp. Phys. **75**(5–6), 331–347 (1989)
12. W. Zwerger, Quantum effects in the current-voltage characteristic of a small Josephson junction. Phys. Rev. B **35**(10), 4737–4753 (1987)
13. H.S. Van der Zant, W.J. Elion, L.J. Geerligs, J.E. Mooij, Quantum phase transitions in two dimensions: experiments in Josephson-junction arrays. Phys. Rev. B **54**(14), 10081–10093 (1996)
14. V. Jaccarino, M. Peter, Ultra-high-field superconductivity. Phys. Rev. Lett. **9**(7), 7–9 (1962)
15. Ø. Fischer, Properties of high field superconductors containing localized magnetic moments. Helv. Phys. Acta **45**, 331–397 (1972)
16. H.W. Meul, C. Rossel, M. Decroux, Ø. Fischer, G. Remenyi, A. Briggs, Observation of magnetic-field-induced superconductivity. Phys. Rev. Lett. **53**(5), 497–500 (1984)
17. L. Balicas, J.S. Brooks, K. Storr, S. Uji, M. Tokumoto, H. Tanaka, H. Kobayashi, A. Kobayashi, V. Barzykin, L.P. Gor'kov, Superconductivity in an organic insulator at very high magnetic fields. Phys. Rev. Lett. **87**(6), 067002 (2001)
18. P. Fulde, R.A. Ferrell, Superconductivity in a strong spin-exchange field. Phys. Rev. **135**(3A), A550–A563 (1964)
19. A.I. Larkin, Y.N. Ovchinnikov, Inhomogeneous state of superconductors. Zh. Eksp. Teor. Fiz. **20**(3), 762–769 (1965)
20. I.J. Lee, A.P. Hope, M.J. Leone, M.J. Naughton, Revisiting the superconducting phase diagram of (TMTSF)2ClO4. Synth. Met. **70**, 747–750 (1995)
21. C.A.R. Sa de Melo, Reentrant superconductivity in quasi-low-dimensional systems : magnetic field effects. J. Supercond. **12**(3), 459–468 (1999)
22. A.G. Lebed, Hidden reentrant and Larkin-Ovchinnikov-Fulde-Ferrell superconducting phases in a magnetic field in a (TMTSF)2ClO4. Phys. Rev. Lett. **107**(8), 087004 (2011)
23. H.A. Radovan, N.A. Fortune, T.P. Murphy, Magnetic enhancement of superconductivity from electron spin domains. Nature **425**, 51–55 (2003)
24. Y. Matsuda, H. Shimahara, Fulde-Ferrell-Larkin-Ovchinnikov state in heavy fermion superconductors. J. Phys. Soc. Jpn. **76**(5), 051005 (2007)

25. V. Zdravkov, A. Sidorenko, G. Obermeier, S. Gsell, M. Schreck, C. Müller, S. Horn, R. Tidecks, L. Tagirov, Reentrant superconductivity in Nb/Cu1-xNix bilayers. Phys. Rev. Lett. **97**(5), 057004 (2006)
26. K. Michaeli, A.C. Potter, P.A. Lee, Superconducting and ferromagnetic phases in SrTiO3/LaAlO3 oxide interface structures: possibility of finite momentum pairing. Phys. Rev. Lett. **108**(11), 117003 (2012)
27. M.Y. Kharitonov, M.V. Feigel'man, Enhancement of superconductivity in disordered films by parallel magnetic field. J. Exp. Theor. Phys. Lett. **85**(7), 421–425 (2005)
28. A. Rogachev, T.C. Wei, D. Pekker, A.T. Bollinger, P.M. Goldbart, A. Bezryadin, Magnetic-field enhancement of superconductivity in ultranarrow wires. Phys. Rev. Lett. **97**(13), 137001 (2006)
29. R. Córdoba, T.I. Baturina, J. Sesé, A.Y. Mironov, J.M. De Teresa, M.R. Ibarra, D.A. Nasimov, A.K. Gutakovskii, A.V. Latyshev, I. Guillamón, H. Suderow, S. Vieira, M.R. Baklanov, J.J. Palacios, V.M. Vinokur, Magnetic field-induced dissipation-free state in superconducting nanostructures. Nat. Commun. **4**, 1437 (2013)
30. F. Lévy, I. Sheikin, B. Grenier, A.D. Huxley, Magnetic field-induced superconductivity in the ferromagnet URhGe. Science **309**(5739), 1343–1346 (2005)
31. E.A. Yelland, J.M. Barraclough, W. Wang, K.V. Kamenev, A.D. Huxley, High-field superconductivity at an electronic topological transition in URhGe. Nat. Phys. **7**(11), 890–894 (2011)
32. Z. Han, A. Allain, H. Arjmandi-Tash, Collapse of superconductivity in a hybrid tin-graphene Josephson junction array. Nat. Phys. **10**, 380–386 (2014)
33. B. Abeles, Effect of charging energy on superconductivity in granular metal films. Phys. Rev. B **15**(5), 2828–2829 (1977)
34. K. Maki, Effect of Pauli paramagnetism on magnetic properties of high-field superconductors. Phys. Rev. **148**(1), 362–369 (1966)
35. N.R. Werthamer, E. Helfand, P.C. Hohenberg, Temperature and purity dependence of the superconducting critical field, Hc2. III. Electron spin and spin-orbit effects. Phys. Rev. **147**(1), 295–302 (1966)
36. A.P. Petrović, D. Ansermet, D. Chernyshov, M. Hoesch, D. Salloum, P. Gougeon, M. Potel, L. Boeri, C. Panagopoulos, A disorder-enhanced quasi-one-dimensional superconductor. Nat. Commun. **7**, 12262 (2016)
37. J. Bardeen, Critical fields and currents in superconductors. Rev. Modern Phys. **34**(4), 667–681 (1962)
38. N. Hadacek, M. Sanquer, J.-C. Villégier, Double reentrant superconductor-insulator transition in thin TiN films. Phys. Rev. B **69**(2), 024505 (2004)
39. A. Gerber, T. Grenet, M. Cyrot, J. Beille, Double-peak superconducting transition in granular LM-Cu-O (L= Pr, Nd, Sm, Eu, D; M= Ce, Th) superconductors. Phys. Rev. Lett. **65**(25), 3201–3205 (1990)
40. K. Parendo, K. Tan, A. Goldman, Evolution towards superconductivity in granular films of bismuth. Phys. Rev. B **76**(10), 100508 (2007)
41. B. Spivak, F. Zhou, Mesoscopic effects in disordered superconductors near Hc2. Phys. Rev. Lett. **74**(14), 2800–2803 (1995)
42. N. Dupuis, G. Montambaux, Magnetic-field-induced Anderson localization in a strongly anisotropic conductor. Phys. Rev. Lett. **68**(3), 357–360 (1992)
43. M. Kunchur, Y.Z. Zhang, P. Lindenfeld, W.L. McLean, J.S. Brooks, Quasireentrant superconductivity near the metal-insulator transition of granular aluminium. Phys. Rev. B **36**(7), 4062–4065 (1987)
44. R. Brusetti, A. Briggs, O. Laborde, M. Potel, P. Gougeon, Superconducting and dielectric instabilities in Tl2Mo6Se6: unusual transport properties and unsaturating critical field. Phys. Rev. B **49**(13), 8931–8943 (1994)
45. A.P. Petrović, R. Lortz, G. Santi, M. Decroux, Phonon mode spectroscopy, electron-phonon coupling, and the metal-insulator transition in quasi-one-dimensional M2Mo6Se6. Phys. Rev. B **82**(23), 235128 (2010)

46. J.M. Kosterlitz, D.J. Thouless, Ordering, metastability and phase transitions in two-dimensional systems. J. Phys. C Solid State Phys. **6**(7), 1181–1203 (1973)
47. M.E. Gershenson, Y.B. Khavin, A.G. Mikhalchuk, H.M. Bolzer, A.L. Bogdanov, Crossover from weak to strong localization in quasi-one-dimensional conductors. Phys. Rev. Lett. **79**(4), 725–728 (1997)
48. D.H. Douglass Jr., Magnetic field dependence of the superconducting energy gap. Phys. Rev. Lett. **6**(7), 346–348 (1961)
49. M. Tinkham, *Introduction to Superconductivity*, 2nd edn. (Dover Publications, USA, 1996)
50. V.K.S. Shante, C.M. Varma, A.N. Bloch, Hopping conductivity in one-dimensional disordered compounds. Phys. Rev. B **8**(10), 4885–4889 (1973)
51. L.W. Gruenberg, L. Gunther, Fulde-Ferrell effect in type-II superconductors. Phys. Rev. Lett. **16**(22), 996–998 (1966)
52. Y. Wang, J.J. Santiago-Avilés, Large negative magnetoresistance and strong localization in highly disordered electrospun pregraphitic carbon nanofiber. Appl. Phys. Lett. **89**(12), 91–93 (2006)
53. V.Z. Kresin, Y.N. Ovchinnikov, Shell structure and strengthening of superconducting pair correlation in nanoclusters. Phys. Rev. B **74**(2), 1–11 (2006)
54. A.A. Shanenko, M.D. Croitoru, Shape resonances in the superconducting order parameter of ultrathin nanowires. Phys. Rev. B **73**(1), 73–76 (2006)
55. S. Bose, A.M. García-García, M.M. Ugeda, J.D. Urbina, C.H. Michaelis, I. Brihuega, K. Kern, Observation of shell effects in superconducting nanoparticles of Sn. Nat. Mater. **9**(7), 550–554 (2010)
56. M. Tian, J. Wang, W. Ning, T.E. Mallouk, M.H.W. Chan, Surface superconductivity in thin cylindrical Bi nanowire. Nano Lett. **15**(3), 1487–1492 (2015)

Chapter 8
Summary and Outlook

8.1 Summary

The motivation for this work rose from the desire to, firstly, improve our understanding of the superconducting transition in low-dimensional materials, and secondly, explore the relationship between superconductivity, electronic correlations, disorder, and dimensionality. The material selected for this purpose was, to the best of our knowledge, the most anisotropic superconductor ever discovered: single crystals of $Na_{2-\delta}Mo_6Se_6$, where δ represents vacancies inherent to the crystal synthesis. The anisotropy arises from the q1D nature of the crystals, composed of Na-doped Mo_6Se_6 nanofilaments weakly coupled transversely to their main axis. Disorder originates from the Na vacancies, hence a lack of electron doping of the nanofilaments and a perturbation in the crystalline field and structure. As a result, the system is an excellent playground to explore the interplay of low-dimensionality, disorder effects, electronic correlations, and the emergence and stability of a superconducting state.

The experimental study of $Na_{2-\delta}Mo_6Se_6$ was mainly performed via transport measurements in the temperature range between milli-Kelvin and room temperature, and in applied magnetic fields up to 15 T. All transport measurements were repeated in different measurement systems to confirm reproducibility, within the accessible ranges of the instruments. Other experimental data were obtained via magnetization measurements, scanning electron microscopy analysis, and X-ray diffraction characterization.

Data analysis was mainly performed via fitting procedures of the transport data. Different temperature regimes were unveiled in the transport measurements (Fig. 8.1) and were identified to be caused by changes in the electronic structure or nature of the materials. These regimes were confirmed via electronic structure calculations and simulations (electron flow, coupling energies, etc.). All results and interpretations of the data showed cross-compatibility within the different tools, adding credence to the conclusions reported in this work.

This study unveiled the presence of exciting electronic phenomena in $Na_{2-\delta}Mo_6Se_6$ single crystals. In brief, $Na_{2-\delta}Mo_6Se_6$ exhibits an atypical super-

D. Ansermet, *Emergent Superconductivity in Low Dimensions*, Springer Theses,
https://doi.org/10.1007/978-981-13-2941-8_8

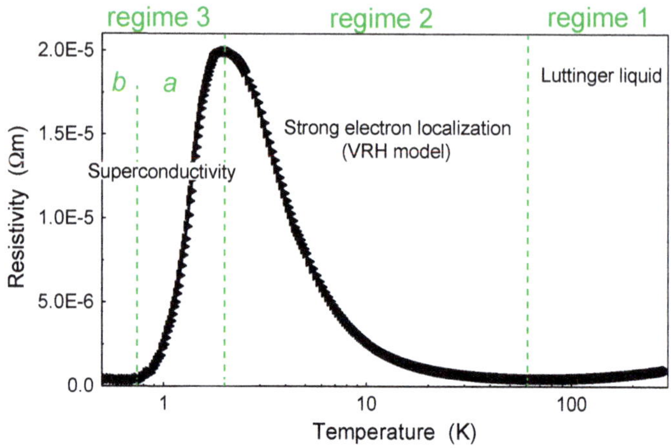

Fig. 8.1 The three different temperature regimes in the transport measurements and the physical phenomenon involved: regime 1 where $dR/dT > 0$ due to Luttinger liquid transport; regime 2 where $dR/dT < 0$ dominated by VRH transport due to strong electron localization; regime 3 where superconductivity develops and stabilizes. The transport data is from sample *B*

conducting ground state emerging from a strongly correlated normal state due to the large anisotropy of the system and the presence of disorder. It is the first time that a superconducting ground state is confirmed in this material, placing it on the top of the list of the most anisotropic superconductors.

Let us walk through the reported and analyzed observations, starting from the normal state (Fig. 8.1).

First, the normal state shows characteristics of 1D electron flow at high temperature, described by Luttinger liquid theory (regime 1 in Fig. 8.1). A renormalization of the Luttinger model including the effect of disorder yields parameters describing attractive electron-electron interactions within the 1D system.

Second, lowering the temperature induces transverse coupling between the chains: a 1D to anisotropic 3D crossover takes place in the electronic normal state. The observed dimensional crossover temperature corresponds well to the one predicted from the electronic band structure, after renormalization due to electron-electron interaction and disorder.

Third, the combination of reduced temperature and intrinsic disorder initiate strong electron localization, where the transport is dominated by electron hopping between localized states (regime 2 in Fig. 8.1). This is described by variable range hopping fits to the experimental transport data and is strongly supported by signatures in the frequency-dependent conductivity, magnetoresistance, and magnetic susceptibility. An observation of a step-like feature emerging between the different parameters extracted for the 6 crystals indicates the presence a mobility edge. This suggests that this q1D material exhibits signatures of 3D electronic transport while remaining highly anisotropic.

Now, let us turn to the emergence and stability of the superconducting ground state (regime 3 in Fig. 8.1).

First, 1D fluctuating superconductivity takes place within the nanofilaments. This is described by fits of thermally activated phase slip and quantum phase slip models to the transport data. In this 1D fluctuating superconducting state, Cooper pairs are confined within 1D filaments and only single electrons can hop between the filaments. At this stage, the normal state is anisotropic 3D while the superconducting state is 1D (zone 3*a* in Fig. 8.1).

Second, a dimensional crossover in the superconducting state takes place via Josephson coupling, once Cooper pairs can tunnel between filaments. This coupling establishes phase coherence in the superconducting state, as phase slips are now destabilized. A coherent state is confirmed by the observation of a small Meissner fraction in the magnetization measurements, the vanishing of the resistance by several orders of magnitude in the transport data, and the emergence of a hump in the superconducting transition. Interestingly, the dimensional crossover can be fitted by the BKT model, suggesting that q1D materials can also exhibit BKT-like transitions in the 2D plane transverse to the chains.

Third, the Josephson coupling responsible for the establishment of phase coherence is strongly dependent on temperature, magnetic field, and current. A detailed study of the superconducting state unveils a reentrant behavior, causing the phase coherence to be stabilized at non-zero temperature, magnetic field or current (zone 3*b* in Fig. 8.1). These experimental observations are successfully described by a simulation of the Josephson energy and are caused by the presence of electron localization in the normal state.

Fourth, the onset of superconducting fluctuations and coherence shows a positive correlation with the level of intrinsic disorder. This is observed in the location of the superconducting peak in the transport measurements, the dependence of the transport behavior with applied magnetic fields and the signatures of Josephson coupling in the superconducting transition. The presence of electron localization and suppression of Coulomb repulsion in the normal state are suggested to cause this enhancement of superconductivity, as predicted by recent theoretical works.

The strong correlations between the normal and superconducting states indicate that the observations of reentrant superconductivity and enhancement by disorder could be more general than just in $Na_{2-\delta}Mo_6Se_6$. This material incorporates all the necessary ingredients for such observations, but this should be seen as just the beginning of the journey.

8.1.1 Project Challenges

Several challenges were faced during this study and should be highlighted.

The crystals were stable to Singapore's climate (humidity, heat) but very sensitive to manipulations including tweezers handling, contact fabrication, mechanical shocks, or repeated thermal cycles. This proved quite challenging as the number of available crystals was fixed: no new crystals were fabricated during the period of this study. Any damage on the crystals was therefore critical, as this was constantly reducing the number of available samples.

The literature on $Na_{2-\delta}Mo_6Se_6$ being quite nonexistent, any unexpected observation opened a door to different possible explanations. Further measurements were not always possible due to the restricted number of measurable crystals.

The q1D nature of the crystals implies that transverse transport data would provide powerful information on the system. However, the physical size of the crystals made the transverse contact fabrication extremely challenging. In addition, the crystalline nature of the samples implies that transverse transport might statistically include bits of longitudinal transport, which are highly complicated to distinguish from transversal transport. A detailed transverse transport analysis was therefore abandoned due to the lack of reproducibility of the acquired data.

The material characterization involved extremely sensitive measurements techniques, which were not necessarily available in Singapore. For example, the X-ray characterization was performed at the European synchrotron radiation facilities in Grenoble. This means that not all crystals could be analyzed for obvious practical reasons, and quantitative disorder levels could not be connected to transport measurements. Therefore, a numerical value of the level of disorder has not been obtained in each crystal.

Finally, several theories applying to this system are not yet fully developed. This can hopefully provoke positive reaction from the community: this study introduces an exciting new system presenting unprecedented behaviors.

8.2 Outlook

The novelty of this material and the reported results indicate that a variety of future studies can be stimulated.

From a chemist perspective, new crystals of $Na_{2-\delta}Mo_6Se_6$ could be grown with controlled amounts of Na vacancies. This could lead to a comprehensive set of crystals where the level of disorder would range over the extended-to-localized states spectrum. The observation of a mobility edge indicates that if disorder can efficiently be tuned, controlled properties could be obtained from the same material with a different level of disorder.

From the material scientist perspective, new q1D materials could be engineered, including the same ingredients as the ones found $Na_{2-\delta}Mo_6Se_6$ single crystals: a superconducting ground state, extreme anisotropy, intrinsic disorder and low density of states with a large dielectric background. Also, differentiating the effects of each ingredient on the transport properties may help to understand the extent of the superconductivity enhancement as well as the reentrant behavior.

From an experimental physicist perspective, further experimental studies of $Na_{2-\delta}Mo_6Se_6$ crystals would provide insightful information on these systems.

First, the transverse transport properties hold information about the dimensional crossover process in the normal and superconducting state. However, the first challenge is to successfully fabricate contacts with transverse geometry. A solution could lie in the peeling of a crystal to isolate one strand of $Na_{2-\delta}Mo_6Se_6$ fiber, then fabricating the contacts via e-beam lithography.

Second, measurements of the electronic heat capacity of $Na_{2-\delta}Mo_6Se_6$ could reveal the electron-phonon coupling strength as well as the evolution of the superconducting state as a function of the dimensional crossover. $Na_{2-\delta}Mo_6Se_6$ is expected to exhibit s-wave pairing symmetry, similarly to $Tl_{2-\delta}Mo_6Se_6$ and $In_{2-\delta}Mo_6Se_6$. However, extracting the specific heat in these crystals is highly non-trivial due to their very low density of states.

Third, Performing solid-state nuclear magnetic resonance (NMR) measurements on these crystals would confirm the BCS s-wave symmetry of this material. Indeed, a specific signature known as a Hebel-Slichter peak appears in NMR measurement only in weakly-coupled s-wave superconductors.

Fourth, it would be interesting to track the possible emergence of a FFLO phase at high field. This is supported by the high values of the Maki parameter and the high experimental critical fields.

As an additional note, APRES measurements have been performed by our collaborators on the $M_{2-\delta}Mo_6Se_6$ and will soon be published. These measurements reproduce perfectly the theoretical band structure and confirm the high one-dimensionality of the Fermi surface. These results open the possibilities of more spectroscopy works to come.

Finally, from a theoretical physicist perspective, many new studies can be initiated by this project. The theory of disordered LL is in its infancy and could provide useful information about the normal state transport. The observation of a mobility edge in low-dimensional systems is largely unexplored and these results are unprecedented. Also, a link between the mobility edge and dimensional crossover is suggested in the data, calling for further studies. Furthermore, the role of disorder on the enhancement of superconductivity still needs to be elucidated. The theories of superconductivity enhancement though multifractality were developed only for 3D and 2D systems and need to be adapted to lower dimensions.

There are certainly other outlooks to this study and the ones presented here constitute only a non-exhaustive list. These, however, indicate the fruitful environment which low-dimensional superconductors offer for fundamental research as well as practical applications.